MODULAR MATHEMATICS
Module E: Mechanics 1

By the same authors

MODULAR MATHEMATICS

Module E: Mechanics 1

L. Bostock, B.Sc.

S. Chandler, B.Sc.

Stanley Thornes (Publishers) Ltd

First published in 1994 by Stanley Thornes (Publishers) Ltd, Ellenborough House, Wellington Street, CHELTENHAM GL50 1YW

97 98 99 00 01 / 10 9 8 7 6 5 4 3

A catalogue record for this book is available from the British Library.

ISBN 0–7487–1502–9

Cover photograph reproduced by courtesy of Allsport photo library.

Typeset by Tech-Set, Gateshead, Tyne & Wear.
Printed and bound by Dah Hua Printing Co. Ltd., Hong Kong

CONTENTS

PREFACE

This book is part of a modular mathematics course designed for students who are working for academic qualifications in mathematics beyond GCSE.

Together with Module A this module covers the work necessary for an AS-level subject combining pure maths and mechanics. Other modules will cover the work for a wider variety of options for AS-level and A-level mathematics courses.

The early chapters do not assume that students already know any mechanics. The book starts with work designed to build up, from the study of simple everyday situations, a framework of basic techniques. Further concepts are then added gradually. For some students this early work can be treated as recapitulation and revision.

Inevitably some of the pure mathematics covered in Module A is used in this book but it is not necessary to have completed the work in Module A before starting on the mechanics. The order of the topics in this book has been arranged so that the two modules can be studied concurrently, the necessary pure mathematics preceding the areas of mechanics where it is applied.

There are many worked examples and plenty of exercises which start with straightforward questions. At intervals throughout the book there are consolidation sections which contain a summary of the work covered in the preceding chapters and an exercise comprising examination questions. It is not intended that these questions be used immediately a topic has been studied; they are better used later for revision, when confidence and some sophistication of style have been developed.

The problems associated with using simple models for real-life situations are discussed in the final chapter, which also gives some suggestions for practical work.

We are grateful to the following examination boards for permission to reproduce questions from their past examination papers. (Any answers included have not been provided by the examining boards, they are the responsibility of the authors and may not necessarily constitute the only possible solutions.)

University of London Examinations and Assessment Council (ULEAC)
Northern Examinations and Assessment Board (NEAB)
University of Cambridge Local Examinations Syndicate (UCLES)
The Associated Examining Board (AEB)
Welsh Joint Education Committee (WJEC)

L. Bostock
S. Chandler

1994

USEFUL INFORMATION

ABBREVIATIONS

$=$	is equal to	$\Rightarrow$	giving, gives or implies
$\approx$	is approximately equal to	$\equiv$	is equivalent to
+ve	positive	$-$ve	negative

A $\circlearrowright$ taking clockwise moments about an axis through A

B $\circlearrowleft$ taking anticlockwise moments about an axis through B

2sf corrected to 2 significant figures 3dp corrected to 3 decimal places

NOTATION USED IN DIAGRAMS

Force $\longrightarrow$ Acceleration $\longrightarrow\!\!\!>$

Velocity $\longrightarrow\!\!\triangleright$ Dimensions $\longleftrightarrow$

Where components and resultant are shown in one diagram the resultant is denoted by a larger arrow-head e.g. $\longrightarrow\!\!\triangleright$

THE VALUE OF *g*

Throughout this book, unless a different instruction is given, the acceleration due to gravity is taken as 9.8 metres per second per second, i.e. $g = 9.8$

ACCURACY OF ANSWERS

Practical problems rarely have exact answers. Where numerical answers are given they are corrected to two or three significant figures or decimal places, depending on their context.

Answers found from drawn graphs may not even be reliable beyond the first significant figure.

USEFUL TRIGONOMETRY

The Sine Rule for a triangle ABC.

$$\frac{a}{\sin A} = \frac{b}{\sin B} = \frac{c}{\sin C}$$

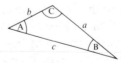

The Intercept Theorem states that a straight line that is parallel to one side of a triangle, divides the other two sides in the same ratio.

In particular, if the parallel line bisects one of the other sides it also bisects the third side. This special case can be called the Midpoint Theorem.

A *median* of a triangle is a line joining a vertex to the midpoint of the opposite side.

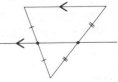

ix

CHAPTER 1

MOTION

DISTANCE, SPEED AND TIME

Mechanics is the study of how and why objects move in various ways, or do not move at all.

Everyone is familiar with time, distance and speed and we begin this book by recalling the relationships between these quantities.

If an object is travelling with constant speed, the distance it covers is given by

$$\textbf{distance} \ = \ \textbf{speed} \times \textbf{time}$$

hence
$$\textbf{speed} \ = \ \frac{\textbf{distance}}{\textbf{time}}$$

Note also that

$$\textbf{average speed} \ = \ \frac{\textbf{total distance}}{\textbf{total time}}$$

When these formulae are used the three quantities must be measured in units that are consistent,

- e.g. if distance is measured in kilometres and time is measured in hours then speed must be measured in kilometres per hour (km/h),
- or if distance is measured in metres and time is measured in seconds then speed must be measured in metres per second (m/s).

DISTANCE–TIME GRAPHS

Suppose that an object is moving in a straight line and that its distances from a fixed point on the line are recorded at various times. By plotting corresponding values, a distance–time graph can be drawn to illustrate the motion of the object.

Consider this situation.

A cyclist travelling along a straight road, covers the 18 km between two points A and B in $1\frac{1}{2}$ hours and the next 35 km, from B to C, in $2\frac{1}{2}$ hours. The graph illustrating this journey is given below.

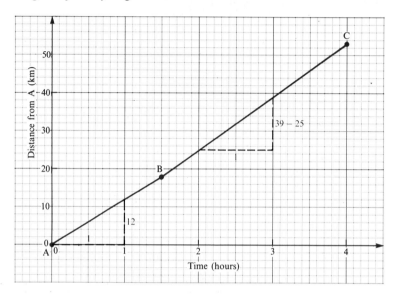

A useful property can be deduced from the graph.

The cyclist's speed from A to B is given by distance ÷ time,

i.e. $18 \div 1\frac{1}{2}$ km/h $= 12$ km/h

The gradient of the graph for the section from A to B is $12 \div 1 = 12$

Also, the cyclist's speed from B to C is $35 \div 2\frac{1}{2}$ km/h $= 14$ km/h

and the gradient of the graph for the section BC is $(39 - 25) \div 1 = 14$

These two results are examples of the general fact that

> **the gradient of the distance–time graph gives the speed**

Curved Distance–Time Graphs

In practice there are many situations where the speed of a moving object is not constant. In such cases the graph of distance plotted against time is not a straight line but a curve.

As an example, consider the distance, d metres, of a car from a set of traffic lights as the car pulls away from the lights.

This table shows the values of d after t seconds.

t	0	1	2	3	4	5
d	0	2	8	18	32	50

The corresponding distance–time graph is:

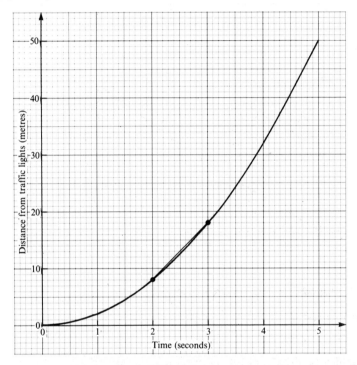

This time the speed of the car cannot be found immediately from the gradient of the graph as the graph is not a straight line. What we can do however is to find the average speed over a chosen interval of time.

Consider, for example, the motion during the third second (i.e. from $t = 2$ to $t = 3$).

From the values in the table, the average speed during this second is

$$\frac{18 - 8}{3 - 2} \text{ m/s} = 10 \text{ m/s}$$

The gradient of the line joining the points on the graph where $t = 2$ and $t = 3$ is also 10.

So the gradient of a line joining two points on a curved distance–time graph gives the average speed in that time interval. (A line joining two points on a curve is called a *chord*.)

Now suppose that we want to find the speed of the car *at the instant* when $t = 2.5$.

We can choose a shorter interval of time, say from $t = 2.2$ to $t = 2.8$ and find the gradient of this chord which is closer to the curve than the first one was. This value gives a better approximation to the speed in the region of $t = 2.5$ and an even better approximation is given by using a yet shorter time interval.

As the ends of the chord get closer to each other the chord becomes nearer and nearer to the tangent to the curve at the point where $t = 2.5$ so we deduce that the speed at the instant when $t = 2.5$ is given by the gradient of the tangent to the curve at this value of t.

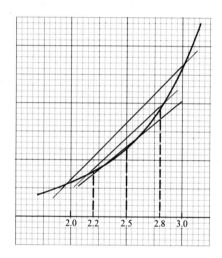

In general

the speed at a particular value of t is given by
the gradient of the tangent to the distance–time graph
at the point where t has that value

Now that this property is established there is no need to find a succession of average speeds by drawing a number of chords; we can go straight to drawing the tangent at the required point. At this stage this is done by judging the position of the tangent visually, so the result can only be a rough approximation to the speed.

EXERCISE 1a

1.

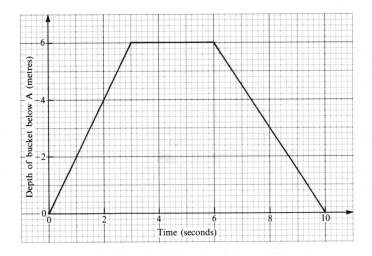

The graph illustrates the motion of a bucket being lowered into a well from the top, A, down to the water level, B, filled with water and drawn up again.

(a) What is the depth of the water level below A?

Find

(b) the speed of the bucket as it descends from A to B

(c) the speed of the bucket as it ascends from B back to A

(d) the time taken to fill the bucket

(e) the average speed for the whole operation including the filling of the bucket.

2. A cyclist rides at 5 m/s along a straight road for 25 minutes. She then dismounts and pushes the bicycle for 5 minutes at 1.7 m/s. Draw a distance–time graph and find the average speed for the whole journey.

3. A goods wagon is shunted 60 m forward in 12 seconds, then 24 m back in 8 seconds and finally 44 m forward in 11 seconds.

(a) Draw the graph of distance from starting point against time.

(b) Write down the speed in each of the three sections of the motion.

(c) Find the average speed for

(i) the first 20 seconds

(ii) the whole journey.

4. In an experiment to measure the viscosity of a lubricant, a ball-bearing was allowed to fall through the lubricant contained in a cylinder. The distance, d centimetres, of the ball-bearing from the bottom of the cylinder was measured after t seconds, for a series of values of t and the results illustrated by this distance–time graph.

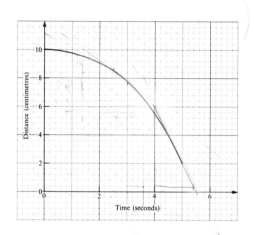

(a) After how long do you think the ball-bearing will reach the bottom?

Find

(b) the average speed during (i) the third second   (ii) the fifth second

(c) the speed when (i) $t = 2.5$ (ii) $t = 4$.

5. A particle P is moving in a straight line from a point A. This table gives the distance, s metres, of P from A, after t seconds.

t (seconds)	0	1	2	3	4	5	6	7	8
s (metres)	0	0.2	0.8	1.8	3.2	5.0	7.2	9.8	12.8

Plot the points and draw a distance–time graph. Hence find

(a) the average speed over the first 4 seconds

(b) the speed when (i) $t = 4.5$ (ii) $t = 6$

(c) for how long the distance of P from A is less than 6 m.

6. A point of light moving in a straight line on the screen of an oscilloscope is at a distance s millimetres from O at a time t seconds where $s = \dfrac{20}{t+1}$.

Draw a distance–time graph for values of t from 0 to 5 and use it to find

(a) the speed, in mm/s, when $t = 2.5$

(b) the time when the point of light is 8 mm from O

(c) the average speed of the light during the five seconds.

SPEED–TIME GRAPHS

This graph illustrates the motion of a dog chasing a rabbit. The dog starts off enthusiastically, then tires a little and finally gives up hope.

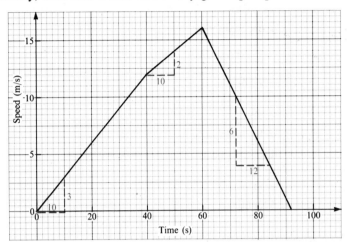

In the early stages of the chase the dog's speed is increasing; we say that the dog is *accelerating*. Acceleration tells us by how much the speed increases in one unit of time and it is measured in a unit of speed per unit of time, e.g. m/s per second, which is written m/s^2. When a speed is going down, the rate at which it decreases is known as *deceleration* or *retardation*.

The following observations can be made from the graph.

(i) In the first section of the motion the dog's speed increases steadily from zero to 12 m/s in 40 seconds, i.e. the dog accelerates at $\frac{12}{40}$ m/s^2 = 0.3 m/s^2. The gradient of the graph for this section is $\frac{3}{10}$ = 0.3.

For the next 20 seconds the dog's speed increases steadily from 12 m/s to 16 m/s, i.e. the dog accelerates at $\frac{2}{10}$ m/s^2 = 0.2 m/s^2.
In this section the gradient of the graph is $\frac{2}{10}$ = 0.2

Finally, in slowing down steadily from 16 m/s to rest in 32 seconds, the dog decelerates at $\frac{16}{32}$ m/s^2 = 0.5 m/s^2
The corresponding gradient is $-\frac{6}{12}$ = -0.5

Each section of the motion illustrates the following general facts.

The gradient of the speed–time graph gives the acceleration;
a negative gradient indicates deceleration

(ii) For the first 40 seconds

the average speed is $\frac{1}{2}(0+12)$ m/s $= 6$ m/s,
the distance covered is therefore 6×40 m $= 240$ m,

the area under the graph (a triangle) is $\frac{1}{2} \times 40 \times 12 = 240$
(using measurements from the scales on the axes).

For the next 20 seconds

the average speed is $\frac{1}{2}(12+16)$ m/s $= 14$ m/s,
the distance covered is therefore 14×20 m $= 280$ m,

the area under the graph (a trapezium) is $\frac{1}{2}(12+16) \times 20 = 280$

For the final 32 seconds

the average speed is $\frac{1}{2}(16+0)$ m/s $= 8$ m/s,
the distance covered is therefore 8×32 m $= 256$ m,

the area under the graph is $\frac{1}{2} \times 32 \times 16 = 256$

In each section the area under the graph represents the distance covered in that section, illustrating this general fact.

The area under a speed–time graph gives the distance covered

Curved Speed–Time Graphs

In the example above the speed changes steadily in each section of the motion so the graph for each section is a straight line. The gradient of that line gives the acceleration during that section of motion and it follows that this acceleration is constant over the section.

On the other hand, the graph of the motion of an object whose speed changes in a variable way is a curve, so there is no section where the acceleration can be 'read' from a straight line graph. A different way is therefore needed to find the acceleration in this case.

Finding the Acceleration

This graph shows the speed of a roller-coaster as it goes from the top of the first climb to the top of the next rise.

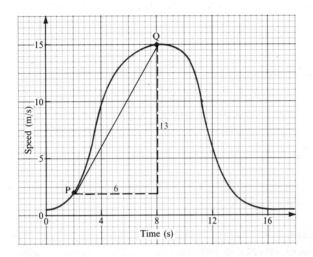

Because the graph of its speed is not made up of straight lines, the acceleration of the roller-coaster at a particular instant cannot be found immediately. However, by using the gradient of a chord joining two points on the graph we can find the *average acceleration* over that part of the motion.

For example the average acceleration, in m/s^2, between the instants when $t = 2$ and $t = 8$ is given by the gradient of PQ,
i.e. the average acceleration is $\frac{13}{6}$ $m/s^2 \approx 2.1$ m/s^2.

When the ends of a chord are brought closer together, the average acceleration gives a better approximation to the actual acceleration within the time interval. Ultimately, when the ends of the chord coincide, the chord becomes a tangent to the curve at a point where t has a particular value, i.e.

**the acceleration at a particular value of t is given by
the gradient of the tangent to the speed–time graph
at the point where t has that value**

Drawing a tangent to a curve is, as we mentioned on page 4, a matter of visual judgement, so an acceleration found in this way is only an approximation.

Note that, to the right of $t = 8$, the gradient of any tangent is negative verifying that the roller-coaster is slowing down, i.e. decelerating.

Finding the Distance Covered

We have seen that if a speed–time graph consists of one or more straight lines, i.e. for each part of the motion the acceleration is constant, the area under the graph gives the distance covered. We showed this, for each separate section of the graph, by using the average speed which, for a straight-line graph, is the speed half-way through the time interval.

When the graph is curved, however, the average speed cannot be found directly in this way, but the method can be adapted to give a good approximation.

If we regard the curve as a series of short straight lines, under each of these lines we have a trapezium whose area can be found.

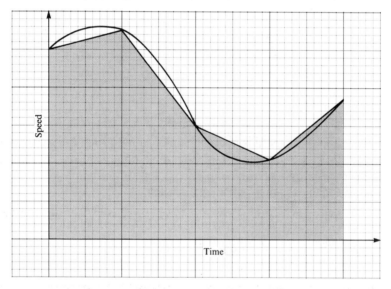

Looking at one of these lines, it is clear that the area underneath it is approximately equal to the area under the corresponding part of the curve. So the sum of the areas beneath all the short lines gives an approximate value for the area under the whole curve.

By making each line successively shorter, so that it gets closer to the curve, the approximation becomes so good that we can now say, whatever the shape of the graph,

**the distance covered in an interval of time
is given by the area under the speed–time graph for that interval**

When dividing a graph into sections to find an approximation to its area, the arithmetic is simpler if strips of equal width are used.

Examples 1b

This graph represents the motion of a cyclist during the first five seconds of a ride.

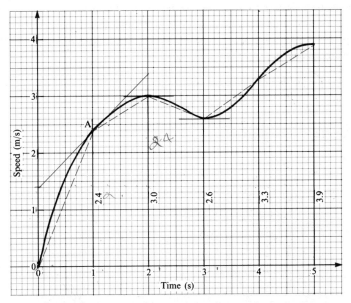

(a) Find an approximate value for the acceleration after 1 second.

(b) For what period of time did the cyclist decelerate?

(c) Find an approximate value for the distance travelled by the cyclist in the five seconds. Is this value an under or over estimate?

(d) At which time(s) was the acceleration zero?

(a) To find the acceleration after 1 second, we draw the tangent at A where $t = 1$.

The gradient of this tangent is approximately $(3.4 - 1.4) \div 2 = 1$
∴ the acceleration after 1 second is about 1 m/s².

(b) When the cyclist decelerates, his speed goes down.

The cyclist decelerates during the third second (from $t = 2$ to $t = 3$).

(c) As the motion spans 5 s, it makes sense to divide the curve into five strips each of width 1 unit. The area of each strip is approximately equal to that of a trapezium.

The area of each strip is approximately $\frac{1}{2}$(sum of parallel sides) × 1
∴ the total area under the curve is given approximately by

$$\tfrac{1}{2}(0 + 2.4) + \tfrac{1}{2}(2.4 + 3) + \tfrac{1}{2}(3 + 2.6) + \tfrac{1}{2}(2.6 + 3.3) + \tfrac{1}{2}(3.3 + 3.9)$$

$$= 13.25$$

Therefore the area under the curve is approximately 13 square units (2 sf is sufficient for an approximation).
The distance covered is approximately 13 metres.

This value is less than the actual value for two reasons:
   (i)   we have rounded the calculated value down
   (ii) the area of the trapeziums is a little less than the area under the curve

(d) The acceleration is zero when the tangent to the curve is horizontal.

The acceleration is zero when $t = 2$ and $t = 3$.

EXERCISE 1b

1.

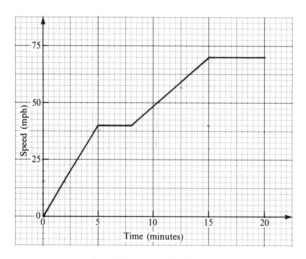

This speed–time graph shows the journey of a train as it moves off from the platform at a station. Find

(a) the acceleration during the first 2 minutes, in mph per minute

(b) the greatest speed of the train

(c) the time for which the train travels at constant speed between its periods of acceleration

(d) the acceleration during the tenth minute.

2. The speed of a motor cycle increases steadily from 12 m/s to 20 m/s in 10 seconds. The rider then immediately brakes and brings the vehicle steadily to rest in 8 seconds. Draw the graph of speed against time for this journey and find

(a) the acceleration

(b) the deceleration

(c) the distance travelled.

3. This graph shows the speed of a bus as it travels between two stops.

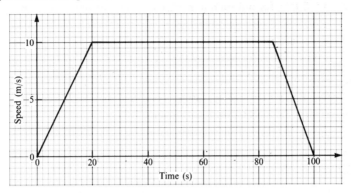

(a) Over what period of time is the bus

(i) accelerating (ii) decelerating (iii) travelling at constant speed?

(b) Find (i) the acceleration (ii) the deceleration.

(c) Find the distance between the two bus stops.

4. A train travelling at 36 m/s starts up an inclined section of track and loses speed steadily at 0.4 m/s². How long will it be before the speed drops to 30 m/s and how far up the incline will the train have travelled by then?

5. This graph shows the speed of a ball that is rolled across a lawn.

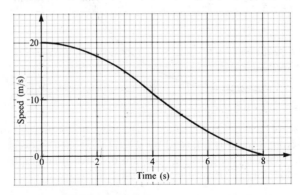

(a) Write down the speed of the ball after

(i) 2 seconds (ii) 7 seconds.

(b) Is the ball accelerating or decelerating when $t = 3$?

(c) Find an approximation for the deceleration during the fifth second.

(d) Find an approximation for the total distance travelled by the ball.

6. A balloon was released into the air on a calm day. Its speed in metres per second was noted at 1 second intervals after release and the results are given in the table.

$t(s)$	0	1	2	3	4	5	6	7	8
$v(m/s)$	0	7	11	13	14	13	11	7	0

Draw the speed–time graph illustrating this data and use it to find

(a) the acceleration of the balloon after (i) 1.5 s (ii) 6 s

(b) the time when the acceleration was zero

(c) the distance travelled by the balloon.

(d) State, with reasons, whether your answer to (c) is an under or over estimate.

7. A rocket is fired and its speed t minutes after firing is v km/minute. This graph shows the corresponding values of t and v during the first 4 seconds of the flight. Find

(a) the acceleration, in km/minute², 3 minutes after firing

(b) the distance covered in the first 3 minutes

(c) the distance covered in the third minute.

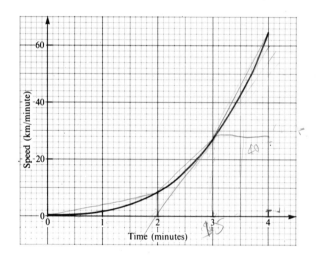

Each question from 8 to 13 is followed by several suggested answers. Choose the correct answer. Some of the questions concern distance–time graphs and others involve speed–time graphs. Read them carefully.

8. This graph shows the motion of an object that starts at O and

A has a constant speed

B is at rest when $t = 4$

C starts with zero speed

D travels a distance of 6 m.

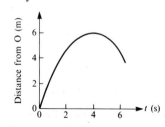

9. A girl throws a ball straight up into the air
and catches it on its way down again. This
graph represents the motion of the ball.

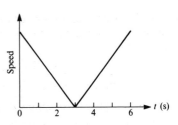

A   The ball goes up with constant speed

B   The acceleration increases as the ball falls

C   The ball is not moving when $t = 3$

D   The distance that the ball rises is negative.

10. A bullet is fired into a block of wood. This graph illustrates the motion of
the bullet inside the block.

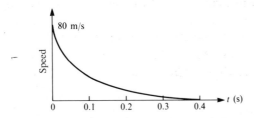

A   The average speed is 40 m/s

B   The bullet has a constant deceleration

C   The bullet stops after 0.4 seconds

D   The bullet penetrates a distance of 16 m into the block.

Use this graph to answer questions 11 and 12.

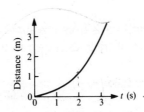

11. The average speed over the 3 seconds of motion is

A $1\frac{1}{2}$ m/s   B   1 m/s   C   3 m/s

12. The speed when $t = 2$ is about

A   0.5 m/s   B   1 m/s   C   2 m/s

13.

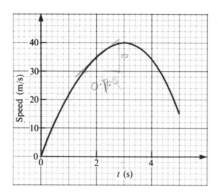

This graph of part of a car journey shows that

A   the car comes to rest when $t = 5$

B   the car changes direction after 3 seconds

C   the acceleration when $t = 2$ is about 1 m/s^2

D   the speed increases for 3 seconds.

MOTION IN A STRAIGHT LINE

If a particle is moving in a straight line it can be moving in either direction along the line. Rather than try to describe these directions in words in every case, we can distinguish between them by giving a positive sign to one of the directions; the other direction is then negative.

DISPLACEMENT

Consider a model engine which starts from a point O and moves in the direction OA along the straight section of track as shown in the diagram.

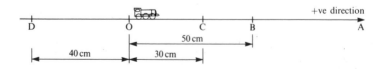

When the engine reaches B, it has travelled a distance of 50 cm and it is also 50 cm from O. However, if the engine then reverses its direction and moves 20 cm back towards O, i.e. to point C,

    the total distance that the engine has travelled is 70 cm

    but the distance of the engine from O is 30 cm (to the right).

If the engine now continues moving towards O and carries on to point D
the total distance that it has travelled is 140 cm
but the distance of the engine from O is 40 cm (to the left).

If we specify the direction from O to A as the positive direction then
the *distance from O in the direction from O to A* is called the *displacement*.

Hence, from O,   the displacement of B is   50 cm

the displacement of C is   30 cm

the displacement of D is −40 cm

Displacement is a quantity which has *both* magnitude (size) *and* direction. Quantities of this type are *vectors*.

Distance, on the other hand, only has size – the direction doesn't matter. Distance is a *scalar* quantity.

Example 1c

Starting from floor 4, a lift stops first at floor 11, then at floor 1 and finally at floor 6. The distance between each floor and the next is 4 m.
Taking the upward direction as positive write down, for each of the stops,

(a) the displacement, s metres, of the lift operator from floor 4

(b) the distance the lift operator has travelled since first leaving floor 4.

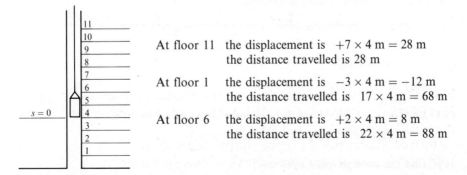

At floor 11   the displacement is $+7 \times 4\,m = 28\,m$
the distance travelled is 28 m

At floor 1   the displacement is $-3 \times 4\,m = -12\,m$
the distance travelled is $17 \times 4\,m = 68\,m$

At floor 6   the displacement is $+2 \times 4\,m = 8\,m$
the distance travelled is $22 \times 4\,m = 88\,m$

EXERCISE 1c

1. Which of the following quantities are vector and which are scalar?

(a) 5 km due south

(e) A temperature of 25°C

(b) 6 miles

(f) A force of 8 units vertically downwards

(c) A speed of 4 m/s

(g) A mass of 6 kg

(d) 200 miles north-east

(h) A time of 7 seconds.

2. A spider is at a point A on a smooth wall, 1.2 m above the floor. This graph represents the motion of the spider as it tries (unsuccessfully!) to climb vertically up the wall.

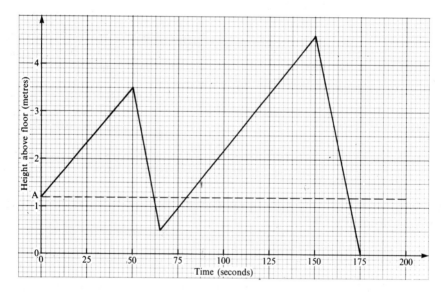

Taking the upward direction as positive,

(a) write down the displacement of the spider from A after

(i)   50 seconds

(ii) 80 seconds

(iii) 175 seconds

(b) find (i)   the maximum displacement of the spider from A

(ii) the total distance the spider travels

(c) find the average speed over the 175 s.

3. A fly is free to move to and fro inside a narrow straight glass tube, closed at both ends. The table below gives the displacement of the fly from the centre of the tube at each instant when it reverses direction. Assuming that the fly moves all the time with a constant speed, illustrate the motion of the fly by drawing a graph of displacement against time.

Time (s)	0	2	5	6	9	10	12
Displacement from O (m)	0	4	−2	0	−6	−4	−8

VELOCITY

If we say that an object is moving with a certain speed *in a particular direction,* we are giving the *velocity* of the object, i.e. velocity is a vector quantity.
Speed is the magnitude of velocity.

Suppose that a body is moving with a *uniform velocity,* i.e. a constant velocity. This means not only that the body has a constant speed, but also that its direction is constant, i.e. that it is moving one way along a straight line.

In order to define which way the body is moving, we specify a velocity in one direction along the line as positive; a velocity in the opposite direction is then negative.

Consider a particle P moving with a constant speed of 5 m/s along the line shown in the diagram.

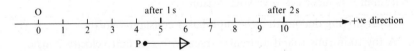

If P starts from O and moves to the right, which is the chosen positive direction, then its velocity is +5 m/s.
After 1 second the displacement of P from O is +5 m
and after 2 seconds it is +10 m,
and so on.

The displacement is increasing at a rate of +5 m/s.

Now if P starts from A and moves to the left with the same speed, its velocity is −5 m/s.

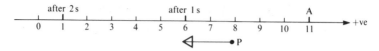

The initial displacement of P from O is +11 m
and after 1 second the displacement is 6 m.

The displacement has decreased by 5 m, i.e. it has increased by −5 m.

Similarly, after 2 seconds the displacement is 1 m and has increased by −10 m.

The displacement is increasing at a rate of −5 m/s

i.e.   **velocity is the rate at which the displacement increases**

ACCELERATION

So far we have associated acceleration only with changing speed. Now that we
have defined velocity, however, we must describe acceleration more carefully,

i.e.   **acceleration is the rate at which *velocity* is increasing**

So if the velocity of a particle moving in a straight line increases steadily from
3 m/s to 11 m/s in 4 seconds, the acceleration is $+2$ m/s^2.

On the other hand, if the velocity goes down from 14 m/s to 5 m/s in 3 seconds
(the velocity has increased by -9 m/s) the acceleration is -3 m/s^2.

Further, if the velocity goes from -4 m/s to -10 m/s in 3 seconds (the velocity
has increased by -6 m/s) the acceleration is -2 m/s^2.

EXERCISE 1d

1. Decide whether each of the following statements is correct or incorrect.
   If you think it is incorrect, give your reason.

   (a) A car driving due north at 40 mph has a constant velocity.

   (b) A toy train runs round a circular track with constant velocity 2 m/s.

   (c) A plane flies in a straight path from London to Newcastle so its velocity is
   constant.

2. A particle is moving along the straight line shown in the diagram. It passes
   through A, travels to B, then moves from B to C, from C to D and finally from
   D to E.

   This table gives the value of t at each point,
   where t is the number of seconds that have
   elapsed since the particle first passed through A.

	B	C	D	E
t	5	8	15	19

Find the velocity of the particle, constant in each section, in travelling from

(a) A to B   (b) B to C   (c) C to D   (d) D to E.

(Remember to give magnitude and direction.)

3. The velocity of a particle changes steadily from -5 m/s to -21 m/s in 4 seconds. What is the acceleration?

4. A particle moving in a straight line with a constant acceleration has a velocity u m/s at one instant and t seconds later the velocity is v m/s. Find the acceleration of the particle if

(a) $u = 8$, $v = 2$, $t = 3$ (b) $u = 4$, $v = -11$, $t = 5$.

5. A body moving initially at 5 m/s has a constant acceleration of a m/s^2. After 6 seconds its velocity is v m/s. Find v if

(a) $a = 3$ (b) $a = -2$ (c) $a = 0$.

DISPLACEMENT–TIME GRAPHS

For an object P moving in a straight line, a displacement–time graph shows how the distance of P *in a specified direction from a fixed point* varies with time.

Consider an object P which moves in a straight line, travelling through points O, A, B and C on the line and covering each section at a constant speed.
This table gives the displacement, s centimetres, of each of these points from O, and the time, t seconds after leaving O, when P is at each point.

	O	A	B	C
s	0	10	6	-9
t	0	5	7	12

The displacement–time graph can now be drawn.

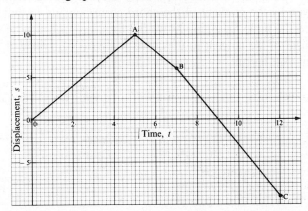

For the section from O to A,
P has travelled a distance of 10 cm in the positive direction in 5 seconds,
i.e. the velocity is 2 cm/s and the gradient of the graph is 2.

For the section from A to B,
P has travelled a distance of 4 cm in the negative direction in 2 seconds,
i.e. the velocity is -2 cm/s and the gradient of the graph is -2.

For the section from B to C,
P has travelled a distance of 15 cm in the negative direction in 5 seconds,
i.e. the velocity is -3 cm/s and the gradient of the graph is -3.

In each case the gradient of the displacement–time graph represents the velocity.

The *average velocity* is the constant velocity that would produce the final increase in displacement in the total time interval, e.g.

the average velocity from O to B is $\frac{6-0}{7-0}$ cm/s $= \frac{6}{7}$ cm/s

 (this is equal to the gradient of the chord OB)

the average velocity from A to C is $\frac{-9-10}{12-5}$ cm/s $= -\frac{19}{7}$ cm/s

 (this is equal to the gradient of the chord AC)

the average velocity from O to C is $\frac{-9-0}{12-0}$ cm/s $= -\frac{3}{4}$ cm/s

 (this is equal to the gradient of the chord OC).

Note that in moving from O to C, the total *distance* that P has moved is

 $(10+4+15)$ cm, i.e. 29 cm.

So P's average *speed* is $29 \div 12$ cm/s, i.e. 2.4 cm/s (2 sf), showing that

 the average speed is *different* from the average velocity.

For a curved displacement–time graph also, the average velocity over a time interval is given by the gradient of the chord corresponding to that interval.
So we can use the same reasoning as we did for a distance–time graph to show that the gradient of the tangent to the curve at a particular value of t, represents the velocity at that instant.

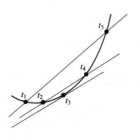

gradient gives
average velocity
from t_1 to t_5

gradient gives
average velocity
from t_2 to t_4

gradient of tangent
gives velocity at t_3

In general, for any type of motion over a given interval of time,

$$\textbf{average speed} \; = \; \frac{\textbf{distance covered in the time interval}}{\textbf{the time interval}}$$

$$\textbf{average velocity} \; = \; \frac{\textbf{increase in displacement over the time interval}}{\textbf{the time interval}}$$

**the velocity at any instant is represented by
the gradient of the displacement–time graph**

Examples 1e

1. A particle P moves along a straight line, starting from a fixed point O on that line. The displacement–time graph for its motion over the first six seconds is given below.

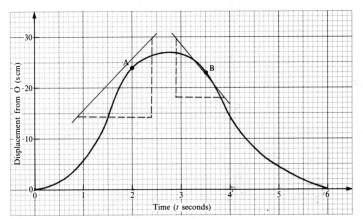

(a)   Find the average velocity from

     (i) $t = 0$ to $t = 2$ (ii) $t = 4$ to $t = 6$ (iii) $t = 0$ to $t = 6$.

(b)   Estimate the velocity of P at the instant when

     (i) $t = 2$ (ii) $t = 3.5$.

(a)   Average velocity is represented by the gradient of a chord.

     (i)   average velocity $= \dfrac{24 - 0}{2 - 0}$ cm/s $= 12$ cm/s

     (ii)   average velocity $= \dfrac{0 - 14}{6 - 4}$ cm/s $= -7$ cm/s

     (iii)   average velocity $= \dfrac{0 - 0}{6 - 0}$ cm/s, i.e. zero

(b) Velocity is represented by the gradient of a tangent.

(i) velocity at A $\approx \dfrac{29 - 13.5}{2.5 - 1}$ cm/s, i.e. 10 cm/s (2 sf)

(ii) velocity at B $\approx \dfrac{17.3 - 29}{4 - 3}$ cm/s, i.e. −12 cm/s (2 sf)

2. A and O are two fixed points on a straight line. A particle P moves on the line so that, at time t seconds its displacement, s metres, from O is given by $s = (t - 1)(t - 5)$. When $t = 0$, the particle is at point A. Draw a displacement–time graph for values of t from 0 to 6.

(a) At what times does the particle pass through O?

(b) What is the average speed over the 6-second time interval?

(c) What is the average velocity over the 6-second time interval?

(d) At what time is the velocity zero?

Using $s = (t - 1)(t - 5)$ for values of t from 0 to 6 gives

t (seconds)	0	1	2	3	4	5	6
s (metres)	5	0	−3	−4	−3	0	5

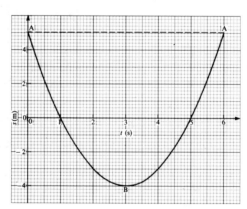

(a) When the particle is at O its displacement from O is zero, i.e. $s = 0$. The particle passes through O when $t = 1$ and $t = 5$.

(b) P starts at A where $s = +5$. P then covers 5 m to O and continues beyond O for a further 4 m to B where $s = -4$. Then P goes back to O, i.e. 4 m back, and a further 5 m to A.

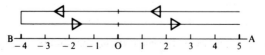

$$\text{average speed} = \frac{\text{total } \textit{distance} \text{ covered}}{\text{time interval}} = \frac{5 + 4 + 4 + 5}{6} = 3 \text{ m/s}.$$

(c)   The average velocity is given by the gradient of the chord joining the points

where $t = 0$ and $t = 6$, i.e. the average velocity is $\dfrac{5-5}{6-0} = 0$

(d)   The velocity is zero when the gradient of the tangent is zero, i.e. when the tangent is parallel to the time axis.

The velocity is zero when $t = 3$.

EXERCISE 1e

1. A boy is practising kicking a football straight towards a wall. For each kick he observes how far the ball rebounds. This graph shows, for one kick, the displacement towards the wall of the ball from the boy.

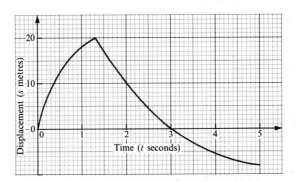

(a)   Use the graph to estimate the velocity of the football
      (i) after half a second   (ii) when $t = 3$ (iii) when $t = 5$.

(b)   (i) State the velocity when $t = 1.3$.

      (ii) Explain why the graph has a 'point' at this time.

(c)   At what time does the football pass the boy when rebounding?

2. A particle P is moving along a straight line. The displacement of P from O, a fixed point on the line, after t seconds is s metres. The table gives some corresponding values of s and t.

t (seconds)	0	1	2	3	4	5	6
s (metres)	0	3	4	3	0	−5	−12

Draw a displacement–time graph and use it to answer the following questions.
(a)   Find the average velocity from
      (i) $t = 0$ to $t = 2$ (ii) $t = 2$ to $t = 4$
      (iii) $t = 0$ to $t = 4$ (iv) $t = 2$ to $t = 6$.

(b)   Find the average speed for each of the time intervals in part (a).

(c)   What do you think the velocity is when $t = 0$?

(d)   At what time is the velocity zero?

3. A is a fixed point on a straight line and P is moving on the line. The displacement, s metres, of P from A after t seconds is given by $s = 5t - t^2$.

(a) Copy and complete the following table.

t	0	1	2	3	4	5	6
s	0		6				

(b) Choose suitable scales and draw a displacement–time graph and use it to answer the following questions.

(c) At what time is the velocity zero?

(d) Estimate the velocity when $t = 5$.

(e) Find, for the 6-second journey,
  (i) the average velocity   (ii) the average speed.

4. This graph illustrates the motion of a ball bouncing vertically.

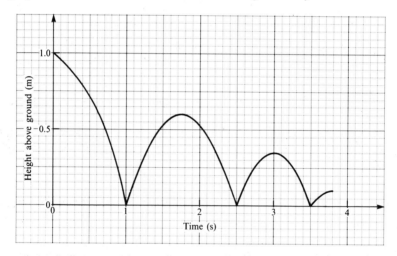

(a) State the times at which the graph shows that the velocity of the ball is zero. (Think carefully before you answer.)

(b) Find the average speed during
  (i) the first second
  (ii) the first full bounce (i.e. from $t = 1$ to $t = 2.5$)
  (iii) the second full bounce.

(c) Find the average velocity during each of the time intervals specified in part (b).

(d) What is the average velocity over the first 3 seconds.

(e) Estimate the velocity of the ball after (i) 0.5 s   (ii) 1.5 s.

VELOCITY–TIME GRAPHS

For an object P travelling in a straight line, a velocity–time graph shows how the speed of P *in a particular direction* varies with time.

This graph shows the variation in the velocity of P during a 20-second period of motion, starting from a point O on the straight line.

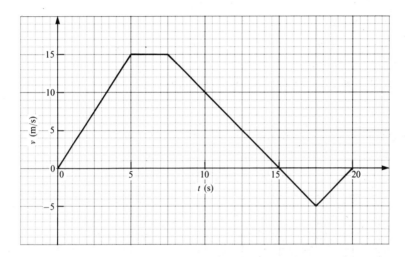

During the first 5 seconds the velocity increases steadily from zero to 15 m/s so the acceleration in this section is 3 m/s^2 and the gradient of this section of the graph is 3.

For the next 2.5 seconds the velocity is constant, i.e. the acceleration is zero, and the gradient the graph is also zero.

Then the velocity decreases until after another 7.5 seconds the graph crosses the time axis. This shows that the velocity has reduced to zero and that P has momentarily come to rest.

For this section the acceleration is $-\frac{15}{7.5}$ m/s^2 and the gradient is -2. Note that, although the velocity is decreasing, it is still positive, so P is still moving forward.

When $t = 15$ the velocity is zero and immediately after that the velocity becomes negative, i.e. P *stops* going forward and begins to move in the opposite direction with an acceleration of $-\frac{5}{2.5}$ m/s^2 (the gradient is -2).

For the last 2.5 seconds the velocity is becoming less negative, i.e. it is increasing. The acceleration is $\frac{0-(-5)}{2.5}$ m/s² $= 2$ m/s² and the gradient is 2. Again note that the velocity is still negative so P is still moving backwards until, when $t = 20$, P comes to rest.

Each section demonstrates that, for motion with constant acceleration, the gradient of the velocity–time graph represents the acceleration.

Now consider the distance moved by P in each section, remembering that P moves forward for 15 seconds and then moves in the reverse direction.

Using average velocity × time gives the following results.

Time interval (s)	0–5	5–7.5	7.5–15	15–17.5	17.5–20
Distance moved (m)	37.5	37.5	56.25	6.25	6.25
Direction moved	fwrd	fwrd	fwrd	bkwrd	bkwrd

As P moves 131.25 m forward and then 12.5 m back, the *displacement* of P from O at the end of the 20 seconds is 118.75 m.

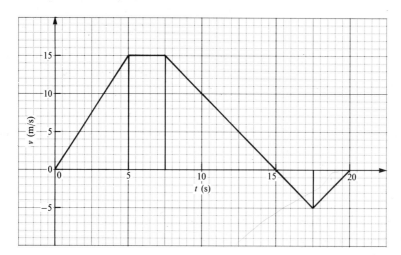

Now we can see that the distance moved in each section is represented by the area between that section of the graph and the time axis. If we take any area that is below this axis as negative, then the displacement of an object moving with constant acceleration is represented by the area between the velocity–time graph and the time axis.

When the acceleration of a moving object is not constant the velocity–time graph is curved. However the arguments we used earlier can be applied again to show that, for a general velocity–time graph:

The average acceleration over a time interval is represented by the gradient of the corresponding chord. A negative gradient gives a negative acceleration, indicating that the velocity is decreasing.

The acceleration at a given instant is represented by the gradient of the tangent to the curve at that particular point on the curve.

The displacement is represented by the area between the curve and the time axis, regions below that axis being negative.
An approximation to this area can be found by dividing it into trapezoidal strips.

EXERCISE 1f

1.

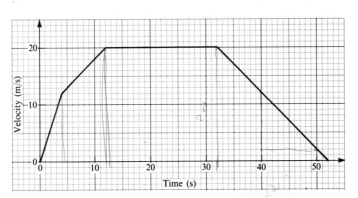

The graph shows the velocity of a car as it moves along a straight road, starting from a lay-by at a point A.

Find

(a) the acceleration during the first 4 seconds

(b) the deceleration during the final 10 seconds

(c) the distance travelled
   (i)   while accelerating
   (ii) at constant speed
   (iii) while decelerating.

Explain why the displacement of the car from A at the end of the 52-second journey is equal to the total *distance* travelled from A.

2. A train is brought to rest from a velocity of 24 m/s with a constant acceleration of −0.8 m/s². Draw a velocity–time graph and find the distance covered by the train while it is decelerating.

3. A particle is moving in a straight line with a velocity of 10 m/s when it is given an acceleration of -2 m/s^2 for 8 seconds. Draw a velocity–time graph for the eight-second time interval and use it to find

(a) the time when the direction of motion of the particle is reversed

(b) the increase in displacement during the 8 seconds

(c) the total distance travelled in this time.

4. The velocity of a runner was recorded at different times and the resulting velocity–time graph is given below.

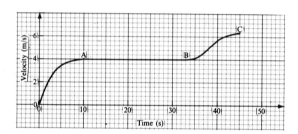

(a) Explain what is happening between
(i) O and A   (ii) A and B   (iii) B and C.

(b) Estimate the length of the race explaining whether your answer is more or less than the actual length.

5. A girl is taking part in an 'It's a Knock Out' game. Starting from her team's base, she has to run forwards to a row of buckets of water, pick one up and run back, trying not to spill any water, to a large cylinder which is at a distance behind the base. Her turn ends when she pours the water into the cylinder. This is the graph of her velocity plotted against time.

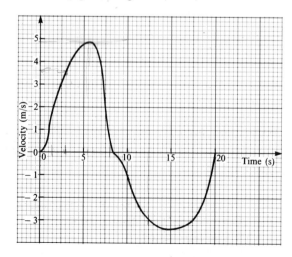

(a) At what time does the girl
      (i) pick up the bucket *8* (ii) empty the bucket? 10

(b) Estimate the acceleration of the girl when $t = 3$ and when $t = 12.5$ $\frac{3-0}{3}=1\cdot2m/s^2$ $\frac{3}{12\cdot5}=0\cdot24m/s^2$

(c) At what times is the acceleration zero? $\frac{3-0}{3}$ = $1\cdot2m/s^2$
      6.5 sec   16 sec

(d) Explain what happens after about 6 seconds.
      6 sec   is there girl decelerates. up to 15 sec and
      over accele.

Each question from 6 to 9 is followed by several suggested answers. State which
is the correct answer.

6. The diagram shows the displacement–time
    graph for a particle moving in a straight line.

    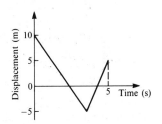

    The average velocity for the interval
    from $t = 0$ to $t = 5$ is

    A   0 .   **B**   6 m/s   **C**   −1 m/s   **D**   2 m/s

7. The diagram shows the displacement–time
    graph for a particle moving in a straight line.

    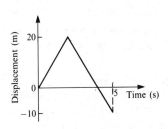

    The distance covered by the particle in the
    interval from $t = 0$ to $t = 5$ is

    A   20 m   **B**   50 m   **C**   15 m   **D**   5 m

8. The diagram shows the velocity–time
    graph for a particle moving in a straight
    line. The sum of the two shaded areas
    represents

    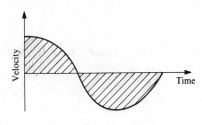

    A   the increase in displacement of the
          particle

    B   the average velocity of the particle

    C   the distance moved by the particle

    D   the average speed of the particle.

9. A particle moving in a straight line with a constant acceleration of 3 m/s² has
    an initial velocity of −1 m/s. Its velocity 2 seconds later is

    A   5 m/s   **B**   6 m/s   **C**   4 m/s   **D**   −7 m/s

    $at + u = v \cdot$
    $3 \times 2 - 1 = v$
    $5 = 5$

CHAPTER 2

CONSTANT ACCELERATION

MOTION WITH CONSTANT ACCELERATION

In Chapter 1 a number of relationships were observed linking the acceleration and velocity of a moving body with the displacement after any time interval.

In the particular case when the acceleration is uniform (i.e. constant) these relationships can be expressed as simple formulae which are known as the equations of motion with constant acceleration.

Consider first the velocity–time graph of an object moving for t seconds with constant acceleration a units.

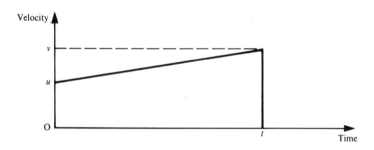

Suppose that at the beginning of the time interval the velocity is u units and at the end it is v units.

The velocity increases by a units each second so after t seconds the increase in velocity is at units.

$$\therefore \qquad\qquad v = u + at \qquad\qquad [1]$$

This formula can be used in solving a problem on motion with uniform acceleration, provided that three out of the four quantities u, v, a and t are known, so that the fourth quantity can be calculated. If this is not the case we need another relationship; this can be found if we consider the displacement, s units, of the object from its starting point after t seconds.

The area under the velocity–time graph represents the displacement. This is the area of a trapezium and is $\frac{1}{2}(u+v) \times t$.

$$\therefore \qquad s = \tfrac{1}{2}(u+v)t \qquad\qquad [2]$$

Now we have a formula to use in those problems where three out of the four quantities u, v, s and t are known.

There are however other possibilities. We could, for example, be given information on the values of u, a and t and have to find the displacement. Neither of the formulae found above link these four quantities but we can use them to deduce another relationship if we eliminate v.

From [1] $\qquad v = u + at$

Substituting in [2] gives $\quad s = \frac{1}{2}(u + u + at)t$

i.e. $\qquad\qquad\qquad s = ut + \tfrac{1}{2}at^2 \qquad\qquad [3]$

In a similar way eliminating u gives

$$s = vt - \tfrac{1}{2}at^2 \qquad\qquad [4]$$

Lastly, a link between u, v, a and s can be found if t is eliminated from [1] and [2].

From [1] $\qquad\qquad\qquad t = \dfrac{v - u}{a}$

Substituting in [2] gives $s = \frac{1}{2}(v + u)\dfrac{(v - u)}{a}$

$$= \frac{1}{2a}(v^2 - u^2)$$

i.e. $\qquad\qquad\qquad v^2 - u^2 = 2as \qquad\qquad [5]$

With these formulae established, we are in a position to tackle, *by calculation*, any problem on motion with constant acceleration.
Each formula contains four quantities, but not the fifth, from u, v, a, s and t so it is easy to identify the one to use by noting which quantity is *not* involved.

However always remember that, as we have already seen, many problems can be solved quickly and easily from a velocity–time graph using only the two basic facts that the gradient gives the acceleration and the area under the graph gives the displacement. Solution by graphical methods should not be neglected because calculation is now an alternative. In fact, even when using the formulae, a velocity–time sketch graph often makes the solution clearer.

Choosing the Positive Direction

Displacement, velocity and acceleration are all vectors and therefore have a direction as well as a magnitude. Because we are considering only motion where the acceleration is constant (its direction is constant as well as its magnitude), it follows that all the motion takes place along a straight line. The object can, however, move either way along the line, so it is necessary to decide which is the positive direction; the opposite direction is then negative.

A good way to start is to make a list of the given information, and what is required, using the standard notation for initial and final velocities, displacement and time, giving each value its correct sign. Most motion problems are made clearer if a simple diagram is drawn, using different arrow heads to indicate different quantities. In this book we use:

$$\longrightarrow\!\!\!\triangleright\qquad\qquad \longrightarrow\!\!\!\!\!\gg\qquad\qquad \longleftarrow\!\!\!\longrightarrow$$

      for velocity,   for acceleration,   for a length

and, later on, $\longrightarrow$ for a force.

Examples 2a

1. A particle starts from a point A with velocity 3 m/s and moves with a constant acceleration of $\frac{1}{2}$ m/s^2 along a straight line AB. It reaches B with a velocity of 5 m/s.

Find (a) the displacement of B from A   (b) the time taken from A to B.

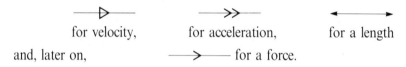

Given: $u = 3$, $v = 5$, $a = \frac{1}{2}$

(a) Required: s

    t is not involved so the formula to use is $v^2 - u^2 = 2as$

$$5^2 - 3^2 = 2(\tfrac{1}{2})s$$
$$\therefore \qquad s = 16$$

The displacement of B from A is 16 m.

(b) Required: t (s is not involved).

    The formula linking u, v, a and t is $v = u + at$

$$5 = 3 + \tfrac{1}{2}t$$
$$\therefore \qquad t = 4$$

The time taken from A to B is 4 seconds.

Note that, as the value of s was found in part (a), in part (b) we could have used the formula linking u, v, s and t, i.e. $s = \frac{1}{2}(u+v)t$, giving $16 = \frac{1}{2}(3+5)t$ and hence $t = \frac{16}{4} = 4$. Remember though that if you use a *calculated* value there is always the risk that it is not correct, making anything found from it wrong as well. Whenever possible it is best to use *given* values.

An alternative method of solution makes use of the velocity–time graph sketched from the given information.

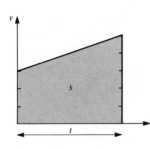

The gradient gives the acceleration.

Gradient $= \dfrac{5-3}{t} = a = \frac{1}{2}$

$\therefore \quad t = 4 \quad \Rightarrow \quad$ the time is 4 seconds.

The area gives the displacement.

Area $= \frac{1}{2}(3+5) \times t = \frac{1}{2}(8)(4) = 16$

$\therefore \quad s = 16 \quad \Rightarrow \quad$ the displacement is 16 m.

2. A cyclist starts riding up a straight steep hill with a velocity of 8 m/s. At the top of the hill, which is 96 m long, the velocity is 4 m/s. Assuming a constant acceleration, find its value.

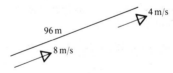

Given: $u = 8$, $v = 4$, $s = 96$

Required: a

The formula without t in it is $v^2 - u^2 = 2as$

$$4^2 - 8^2 = 2a(96)$$

$\Rightarrow \qquad\qquad a = \dfrac{-48}{192} = \dfrac{-1}{4}$

The acceleration is $-\frac{1}{4}$ m/s^2.

Remember that a negative acceleration is called a deceleration (or a retardation) and it indicates that velocity is decreasing.

A sketch of the velocity–time graph shows that this is the case.

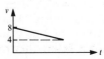

3. The driver of a train begins the approach to a station by applying the brakes to produce a steady deceleration of 0.2 m/s^2 and brings the train to rest at the platform in 1 minute 30 seconds. At the moment when the brakes were applied find

(a) the speed of the train in kilometres/hour

(b) the distance travelled before stopping.

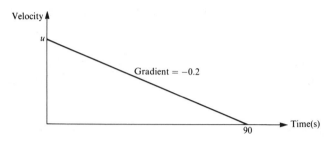

Working in metres and seconds we have:

Given: $a = -0.2$, $v = 0$, $t = 90$

Required: (a) u (b) s

(a)   Using $v = u + at$ gives

$$0 = u + (-0.2)(90)$$

$$\therefore \qquad u = 18$$

The speed of the train was 18 m/s,

i.e. $\dfrac{18}{1000} \times 60 \times 60$ km/h $= 64.8$ km/h.

(b)   Using $s = vt - \frac{1}{2}at^2$ gives

$$s = (0)(90) - \tfrac{1}{2}(-0.2)(90)^2 = 810$$

This is the displacement of the station from the point where the brakes were applied.

$\therefore$ the distance of the train from the station was 810 m.

In a situation where more than one object is involved, or where one object moves in different ways in various sections of the motion, it may be necessary to use 'link' quantities. These are unknown quantities that occur in the motion of more than one object or in more than one section. Whether the problem is tackled by using formulae or by referring to a graph, these quantities enable two incomplete pieces of information to be merged to produce a result.

4. At the same instant two children, who are standing 24 m apart, begin to cycle directly towards each other. James starts from rest at a point A, riding with a constant acceleration of 2 m/s^2 and William rides with a constant speed of 2 m/s. Find how long it is before they meet.

All quantities will be measured using metres and seconds.

James and William start together so they meet after riding for the same time.

For James

    Given: $u = 0$ Involved: time, t

              $a = 2$ distance, s_1

    Using $s = ut + \frac{1}{2}at^2$ gives $s_1 = \frac{1}{2}(2)t^2$ $\Rightarrow$ $s_1 = t^2$

For William

    Given: constant velocity, 2   Involved: time, t

                                              distance, s_2

    Using   distance = velocity × time gives $s_2 = 2t$

The sum of the distances they run is 24 m, i.e. $s_1 + s_2 = 24$

  $\therefore$ $t^2 + 2t = 24$ $\Rightarrow$ $t^2 + 2t - 24 = 0$

                               $\Rightarrow$ $(t + 6)(t - 4) = 0$

                               $\Rightarrow$ $t = -6$ or $t = 4$

Only a positive value for t has any meaning.

The boys meet after 4 seconds.

Note that in this problem the link quantity is t. Neither of the two separate equations gives an actual value of anything, but each gives a different distance in terms of t. These expressions combined are equal to the known total distance.

5. A particle A starts from rest at a point O and moves on a straight line with constant acceleration 2 m/s². At the same instant another particle B, 12 m behind O, is moving with velocity 5 m/s and has a constant acceleration of 3 m/s². How far from O are the particles when B overtakes A?

A and B travel for different distances. We will choose s to represent the shorter distance, i.e. the distance that A travels; B then travels a distance $(s + 12)$. A and B move for the same time.

For A

    Given: $u = 0$, $a = 2$ Involved: distance, s and time, t

            $s = ut + \frac{1}{2}at^2$ $\Rightarrow$ $s = \frac{1}{2}(2)t^2$ [1]

For B

    Given: $u = 5$, $a = 3$ Involved: distance, $(s + 12)$ and time, t

            $s = ut + \frac{1}{2}at^2$ $\Rightarrow$ $s + 12 = 5t + \frac{1}{2}(3)t^2$ [2]

Although we want the distance we will eliminate s first and find t, because t appears to powers 1 and 2 and is difficult to eliminate.

$$[2] - [1] \text{ gives} \quad 12 = 5t + \tfrac{1}{2}t^2 \quad \Rightarrow \quad t^2 + 10t - 24 = 0$$
$$\Rightarrow \quad (t+12)(t-2) = 0$$
$$\Rightarrow \quad t = 2 \text{ or } -12$$

Using the positive value of t in [1] gives $s = 4$.

B overtakes A at a distance 4 m from O.

EXERCISE 2a

In questions 1 to 10 an object is moving with constant acceleration a m/s² along a straight line. The velocity at the initial point O is u m/s and t seconds after passing through O the velocity is v m/s. The displacement from O at this time is s m. For each question select the appropriate formula.

1. $u = 0$, $a = 3$, $v = 15$; find t.
2. $t = 10$, $s = 24$, $u = 6$; find v.
3. $a = 5$, $u = 4$, $s = 2$; find v.
4. $u = 16$, $v = 8$, $t = 5$; find s.
5. $t = 7$, $u = 3$, $v = 17$; find a.
6. $t = 7$, $u = 17$, $v = 3$; find a.
7. $u = 5$, $t = 3$, $a = -2$; find s.
8. $a = -4$, $u = 6$, $v = 0$; find t.
9. $v = 3$, $t = 9$, $a = 2$; find s.
10. $v = 7$, $t = 5$, $a = 3$; find u.

The remaining questions can be solved either by calculation from formulae or by reference to a velocity–time graph.

11. The driver of a car travelling on a motorway at 70 mph suddenly sees that the traffic is stationary at an estimated distance of 60 m ahead. He immediately applies the brakes which cause a deceleration of 6 m/s². Can a collision be avoided? (70 mph $\approx$ 32 m/s)

12. A bowls player projects the jack along the green with a speed of 4 m/s. It comes to rest 'short' at a distance of 25 m. What is the retardation caused by the surface of the green? With what speed should the jack be projected to reach a length of 30 m?

13. A particle is moving in a straight line with a constant acceleration of 2 m/s². If it was initially at rest find the distance covered

    (a) in the first four seconds   (b) in the fourth second of its motion.

14. A particle moving in a straight line with a constant acceleration of -3 m/s^2, has an initial velocity at point A of 10.5 m/s.

  (a) Show that the times when the displacement from A is 15 m are given by $t^2 - 7t + 10 = 0$ and find these times.

  (b) Find the times when the displacement from A is -15 m.

15. A racing car is travelling at 130 mph when the driver sees a broken-down car on the track $\frac{1}{10}$ of a mile ahead. Slamming the brakes on he achieves his maximum deceleration of 24.5 mph per second. How far short of the broken-down car does he stop?

16. A body moving in a straight line with constant acceleration takes 3 seconds and 5 seconds to cover two successive distances of 1 m. Find the acceleration. (Hint: use distances of 1 m and 2 m from the start of the motion.)

17. The displacements from a fixed point O, of an object moving in a straight line with constant acceleration, are 10 and 14 metres at times of 2 and 4 seconds respectively after leaving O. Find

  (a) the initial velocity   (b) the acceleration

  (c) the time interval between leaving O and returning to O.

18. A particle starts from a point O with an initial velocity of 2 m/s and travels along a straight line with constant acceleration 2 m/s^2. Two seconds later another particle starts from rest at O and travels along the same line with an acceleration of 6 m/s^2. Find how far from O the second particle overtakes the first.

19. Starting from rest at one set of traffic lights, a car accelerates from rest to a velocity of 12 m/s. It maintains this speed for 42 seconds, until it decelerates to rest at the next set of red lights 60 seconds after leaving the first set. If the acceleration and deceleration are equal, find the distance between the two sets of lights.

20. A stolen car, travelling at a constant speed of 40 m/s, passes a police car parked in a lay-by. The police car sets off three seconds later, accelerating uniformly at 8 m/s^2. How long is it before the police car intercepts the stolen vehicle and how far from the lay-by does this happen?

21. A particle P, moving along a straight line with constant acceleration 0.3 m/s^2, passes a point A on the line with a velocity of 20 m/s. At the instant when P passes A, a second particle Q is 20 m behind A and moving with velocity 30 m/s. Prove that, unless the motion of P and/or Q changes, the particles will collide.

22. A bus pulls away from a stop with an acceleration of 1.5 m/s^2 which is maintained until the speed reaches 12 m/s. At the same instant a girl who is 5 m away from the bus stop starts to run after the bus at a constant 7 m/s. Will the girl catch the bus?

FREE FALL MOTION UNDER GRAVITY

In the early days of the study of moving objects it was thought that if two objects with different masses were dropped, the heavier one would fall faster than the lighter. This idea was proved to be false by Galileo in a famous series of experiments; it is alleged that he dropped various objects from the top of the leaning tower of Pisa and timed their descents by the Cathedral clock opposite. Whether this anecdote is fact or fiction, the experiments showed that, regardless of their mass, all objects which gave rise to negligible air resistance had the same acceleration vertically downward when falling freely.

This *acceleration due to gravitational attraction* is represented by the letter g. Its value varies marginally in different parts of the world but it is acceptable to take 9.8 m/s^2 as a good approximation. In some circumstances 10 m/s^2 is good enough.

An object falling completely freely travels in a vertical line, so problems on its motion can be solved by using the equations for uniform acceleration in a straight line.

For bodies that are dropped, the downward direction is usually taken as positive, i.e. we take the positive value of g for the acceleration.

If, on the other hand, a ball is thrown vertically upwards, the upward direction could be chosen as positive; in this case the acceleration is $-g$.

Examples 2b

Take the value of g as 9.8 unless otherwise instructed.

1. A brick is dropped from a scaffold board and hits the ground 3 seconds later. Find the height of the scaffold board.

Anything that is 'dropped' from a stationary base is not thrown but released from rest, i.e. its initial velocity is zero.

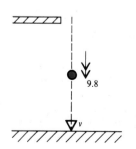

Taking the downward direction as positive:

Known: $u = 0$ Required: s

$a = 9.8$

$t = 3$

Using $s = ut + \frac{1}{2}at^2$ gives

$s = 0 + \frac{1}{2}(9.8)(3^2)$

$\therefore \qquad s = 44.1$

The height of the scaffold is 44.1 m.

2. A boy throws a ball vertically upwards from a seven-metre-high roof.

(a) If, after 2 seconds, he catches the ball on its way down again, with what speed was it thrown?

(b) What is the velocity of the ball when it is caught?

(c) If the boy fails to catch the ball, with what speed will it hit the ground?

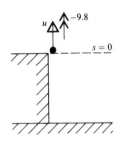

Taking the upward direction as positive:

(a) Known: $t = 2$ Required: u

$a = -9.8$

$s = 0$

Using $s = ut + \frac{1}{2}at^2$ gives

$0 = 2u + \frac{1}{2}(-9.8)(4)$

$\therefore$ $u = 9.8$

The ball was thrown with a speed of 9.8 m/s.

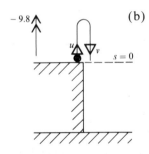

(b) Known: $t = 2$ Required: v

$a = -9.8$

$s = 0$

Using $s = vt - \frac{1}{2}at^2$ gives

$0 = 2v - \frac{1}{2}(-9.8)(4)$

$\therefore$ $v = -9.8$

The ball is *falling* at 9.8 m/s when caught.

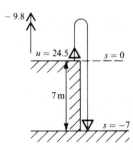

(c) The ground is 7 m below the point of projection so the final displacement of the ball is -7 m.
Working from the time when the ball was thrown we have:

Known: $u = 9.8$ Required: v

$s = -7$

$a = -9.8$

Using $v^2 - u^2 = 2as$ gives

$v^2 - (9.8)^2 = 2(-9.8)(-7)$

$\Rightarrow$ $v^2 = 96.04 + 137.2 = 233.24$

$\therefore$ $v = 15.27\ldots$

The ball hits the ground at 15.3 m/s (3 sf).

Note that in parts (a) and (b) we have shown that a particle thrown upwards with a velocity u, returns to the same level with a velocity $-u$, i.e. with equal speed in the downward direction.

3. One clay pigeon, A, is fired vertically upwards with a speed of 40 m/s and 1 second later another one, B, is projected from the same point with the same velocity. Taking the acceleration due to gravity as 10 m/s², find

(a) the time that elapses before they collide

(b) the height above the point of projection at which they meet.

When B has been in the air for t seconds, A has been in the air for $(t+1)$ seconds. (In a case like this, using t for the shorter time interval and $(t+1)$ for the longer one, rather than t for the first and $(t-1)$ for the second, avoids the algebraic mistakes that can occur when minus signs are involved.)

Taking the upward direction as positive we have:

For A		For B	
Known:	$u = 40$	Known:	$u = 40$
	$a = -10$		$a = -10$
	time $= (t+1)$		time $= t$
Required:	s	Required:	s

(a) Using $s = ut + \frac{1}{2}at^2$ gives:

  For A $s = 40(t+1) + \frac{1}{2}(-10)(t+1)^2$

  For B $s = 40t + \frac{1}{2}(-10)t^2$

  $\therefore$ $40(t+1) - 5(t+1)^2 = 40t - 5t^2$

  $\Rightarrow$ $40 - 10t - 5 = 0$

  $\Rightarrow$ $t = 3.5$

The projectiles collide 3.5 s after B is fired.

(b) As A and B are at the same height when they collide we can find s for either of them.

Considering B,

$$s = 40(3.5) - 5(3.5)^2 = 78.75$$

A and B collide 78.8 m above the point of projection (3 sf).

4. A missile is fired vertically upwards with speed 47 m/s. Find, to 3 sf, the time that elapses before the missile returns to the firing position if it is projected from a point on

(a) a seashore where g can be taken as 9.819

(b) a high mountain where g can be taken as 9.783

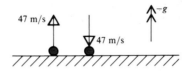

We know that the missile will return with a speed of 47 m/s downwards. Taking the upward direction as positive, we have in either case,

> Known: $u = 47$ Required: t
> $v = -47$
> the value of g

(a) Using $v = u + at$ with $a = -9.819$ gives
$$-47 = 47 + (-9.819)t \quad \Rightarrow \quad t = 9.573\ldots$$

The time of flight of the missile is 9.57 s (3 sf)

(b) Again using $v = u + at$ but with $a = -9.783$ gives
$$-47 = 47 + (-9.783)t \quad \Rightarrow \quad t = 9.608\ldots$$

The time of flight of the missile is 9.61 s (3 sf)

5. A youth is playing with a ball in a garden surrounded by a wall 2.5 m high and kicks the ball vertically up from a height of 0.4 m with a speed of 14 m/s. For how long is the ball above the height of the wall? Give answers corrected to 2 significant figures.

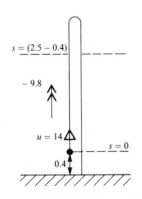

We will measure displacement upward from a point 0.4 m above the ground.

> Known: $u = 14$ Required: t
> $a = -9.8$
> $s = 2.1$

Using $s = ut + \frac{1}{2}at^2$ gives
$$2.1 = 14t - 4.9t^2$$
$$\Rightarrow \quad 0.3 = 2t - 0.7t^2$$
$$\Rightarrow \quad 7t^2 - 20t + 3 = 0$$

t is found from this quadratic equation by using the formula.

i.e. $t = \dfrac{20 \pm \sqrt{(400 - 84)}}{14} = 2.698 \text{ or } 0.159$

The ball is at the height of the top of the wall at two different times. Therefore it takes 0.159 seconds to reach the top of the wall when going up, and returns to that height 2.698 seconds from the start. So the ball is above the wall for 2.5 s, corrected to 2 sf.

EXERCISE 2b

Unless another instruction is given, take g as 9.8 and give answers corrected to 3 significant figures.

1. A stone dropped from the top of a cliff takes 5 seconds to reach the beach.

   (a) Find the height of the cliff.

   (b) With what velocity would the stone have to be thrown vertically downward from the top of the cliff, to land on the beach after 4 seconds?

2. A particle is projected vertically upward from ground level with a speed of 24.5 m/s. Find

   (a) the greatest height reached

   (b) the time that elapses before the particle returns to the ground.

3. A slate falls from the roof of a high-rise building. Find how far it falls

   (a) in the first second

   (b) in the first two seconds

   (c) during the third second.

4. A ball is thrown vertically upward and is caught at the same height 3 seconds later. Find

   (a) the distance it rose

   (b) the speed with which it was thrown.

5. A brick is dropped down a disused well, 50 m deep.

   (a) For how long does it fall?

   (b) With what speed does it hit the bottom?

6. A boulder slips from the top of a precipice and falls vertically downwards on to a plain 200 m below.

   (a) Find, to three significant figures, the speed of the boulder when it hits the plain if the precipice is
      (i) in a polar region where, to four significant figures, the acceleration due to gravity is 9.830 m/s^2
      (ii) in a region near to the equator where the acceleration due to gravity is 9.781 m/s^2 (4 sf)

   (b) Can you find a possible explanation for the difference in the values of g at the two locations (you may need to refer to an encyclopedia).

7. A parachutist is descending vertically at a steady speed of 2 m/s when his watch strap breaks and the watch falls. If the watch hits the ground 3 seconds later at what height was the parachutist when he dropped it?

8. A mine inspector has run into trouble 30 m down an open vertical shaft. To summon help he fires a distress flare straight up the shaft. In order for the flare to be seen, it must reach at least 10 m above the ground level. What is the least speed with which it must be fired?

9. A ball is thrown vertically, with a speed of 7 m/s from a balcony 14 m above the ground. Find how long it takes to reach the ground if it is thrown

   (a) downwards   (b) upwards.

   Find also the speed with which it reaches the ground in each of these cases.

10. A stone is dropped from the top of a building and at the same instant another stone is thrown vertically upward from the ground below at a speed of 15 m/s. If the stones pass each other after 1.2 seconds find the height of the building.

11. A youth playing in a yard surrounded by a 3 metre-high wall kicks a football vertically upward with initial speed 15 m/s.

    (a) What is the greatest height reached by the ball?

    (b) For how long can it be seen by someone on the other side of the wall?

12. A small ball is released from a point 1.6 m above the floor. When it hits the floor its speed is halved. How high does it bounce?

13. A stone is dropped from the top of a building to the ground. During the last second of its fall it moves through a distance which is $\frac{1}{5}$ of the height of the building. How high is the building?

14. A competitor is attempting a dive from a springboard that is 6 m above the water. He leaves the springboard with an upward velocity of 7 m/s. Taking the value of g as 10, find the speed at which the diver enters the water and the time for which he is in the air.

15. The defenders of Castle Dracula dropped large rocks from the battlements on to an attacking army. If the height of the battlements above the ground was 35 cubits find, in cubits per second, the speed at which the rocks that missed their targets hit the ground. Take the value of g as 20 cubits per second per second. (A cubit is an ancient measure of length based on a man's forearm and is approximately half a metre.)

CHAPTER 3

VECTORS

DEFINING A VECTOR

A vector quantity is one for which direction is important as well as magnitude (i.e. size). We have already met, in Chapter 1, some important vector quantities.

Displacement is a distance measured in a particular direction, e.g.
　　　'10 miles due north' is a vector,
whereas '10 miles' is a distance with no specified direction so is not a vector.
A quantity that possesses magnitude only is a *scalar* quantity.
Distance is the magnitude of displacement.

Velocity includes both speed and direction of motion, so
　　　'150 km/h on a bearing of 154°' is a velocity and is a vector,
whereas 'a speed of 150 km/h' is a scalar.
Speed is the magnitude of velocity.

Acceleration is the rate at which velocity is increasing so it follows that acceleration depends on both the speed and direction of motion, i.e. acceleration is a vector.
(Note that there is no different word for the magnitude of acceleration.)

Force is another quantity which plays an important part in the study of mechanics. Clearly if a force pushes an object we need to know both the size of the push and also which way it is acting, i.e. force is a vector.
Force is measured in newtons (N).

VECTOR REPRESENTATION

Any vector can be represented by a line (called a *line segment*). The direction of the line gives the direction of the vector and the length of the line represents the magnitude of the vector.

If the line is labelled $\underline{AB}$, then the vector it represents is written $\overrightarrow{AB}$.
The order of the letters indicates the direction of the vector.
(The vector $\overrightarrow{BA}$ is represented by the same line but in the opposite direction.)

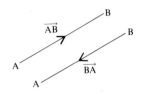

Alternatively a single lower case letter in the middle of the line can be used. In this case there *must* be an arrow on the line to show the direction of the vector.
In print the letter is set in bold type, e.g. **a**.
For hand-written work use $\underline{a}$.
The magnitude of **a** is written as a

Related Vectors

Two vectors are *equal* if they have equal magnitudes *and* the same direction.

We write $\mathbf{a} = \mathbf{b}$

If the direction of **b** is reversed then **a** and **b** are *equal and opposite*.

We write $\mathbf{a} = -\mathbf{b}$

If two vectors have the same direction but different magnitudes then one can be expressed as a multiple of the other, e.g.

$$\mathbf{b} = 2\mathbf{a} \quad \text{and} \quad \mathbf{q} = 3\mathbf{p}$$

In general, if **a** and **b** are parallel then

$$\mathbf{a} = k\mathbf{b} \quad \text{where } k \text{ is a constant of proportion and is scalar.}$$

For example, if $\overrightarrow{AB}$ represents a vector **p**,
then $\frac{1}{3}\mathbf{p}$ is represented by $\overrightarrow{AC}$
where $AC = \frac{1}{3}AB$ and $2\mathbf{p}$ is represented
by $\overrightarrow{AD}$ where AB is extended so that
$AD = 2AB$.

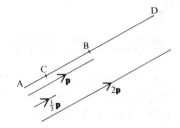

ADDITION OF VECTORS

Consider what happens if a hiker starts from the corner A of a field, walks for 30 m beside the hedge along one side to B and then 40 m along the side perpendicular to the first, to C. The hiker could have reached the same point C by walking directly across the field (assuming this to be allowed!).

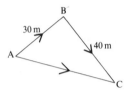

So combining the two displacements $\overrightarrow{AB}$ and $\overrightarrow{BC}$ gives the same final result as the single displacement $\overrightarrow{AC}$.

This is what is meant by *adding vectors* and we can write

$$\overrightarrow{AB} + \overrightarrow{BC} = \overrightarrow{AC}$$

or $\qquad \mathbf{a} + \mathbf{b} = \mathbf{c}$

$\overrightarrow{AC}$ is called the *resultant* of $\overrightarrow{AB}$ and $\overrightarrow{BC}$.

Note that, in this context, + means 'together with' or 'followed by'
and = means 'is equivalent to'.

Note also that, from A we can go to C either directly or via the vectors $\overrightarrow{AB}$ and $\overrightarrow{BC}$. The first point and last point are the same in both cases. Triangle ABC is known as a *triangle of vectors* and when we use it to add vectors we are using the *triangle law*.

Using the Triangle Law

Given two vectors **a** and **b**, represented by line segments $\overrightarrow{OA}$ and $\overrightarrow{OB}$, we can draw diagrams to represent various combinations of **a** and **b**.

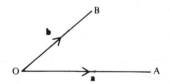

Now **a** + **b** means **a** *followed by* **b**. We can represent this by drawing AC equal and parallel to OB. Then $\overrightarrow{OC}$ represents **a** + **b**, i.e.

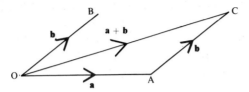

Note that we could equally well have drawn **b** followed by **a**.

To represent **a** − **b** we can draw **a** followed by −**b**.

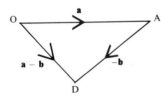

Alternatively we can draw −**b** followed by **a**.

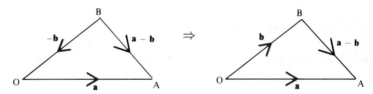

Any number of vectors can be added by the same process, e.g. **a** + **b** + **c** + **d** means **a** followed by **b** followed by **c** followed by **d** and can be represented by $\overrightarrow{OA}$, $\overrightarrow{AB}$, $\overrightarrow{BC}$ and $\overrightarrow{CD}$ as shown.

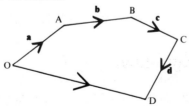

From the diagram we see that $\overrightarrow{OD}$ is equivalent to $\overrightarrow{OA} + \overrightarrow{AB} + \overrightarrow{BC} + \overrightarrow{CD}$

i.e. $\overrightarrow{OD} = \mathbf{a} + \mathbf{b} + \mathbf{c} + \mathbf{d}$

So $\overrightarrow{OD}$ is the resultant of **a**, **b**, **c** and **d**

Note that the arrow we use for marking a resultant vector is larger than those used for the vectors being added.

Examples 3a

1.

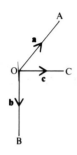

Given the vectors **a**, **b** and **c**, represented by OA, OB and OC as shown, sketch a diagram to illustrate $\mathbf{a} + \mathbf{b} + \mathbf{c}$.

a + **b** + **c** means **a** followed by **b** followed by **c** so we draw OA followed by a line AD, equal and parallel to OB, followed by a line DE, equal and parallel to OC.

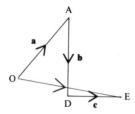

$\overrightarrow{OE}$ represents $\mathbf{a} + \mathbf{b} + \mathbf{c}$.

2. In a triangle ABC, M is the midpoint of AC. $\overrightarrow{AB} = \mathbf{a}$ and $\overrightarrow{BC} = \mathbf{b}$.
Find in terms of **a** and **b**, (a) $\overrightarrow{AC}$ (b) $\overrightarrow{CA}$ (c) $\overrightarrow{AM}$ (d) $\overrightarrow{MB}$.

In the diagram, $\overrightarrow{AC}$ is equivalent to $\overrightarrow{AB}$ followed by $\overrightarrow{BC}$.

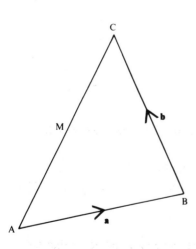

(a) $\overrightarrow{AC} = \overrightarrow{AB} + \overrightarrow{BC}$
$= \mathbf{a} + \mathbf{b}$

(b) $\overrightarrow{CA} = -\overrightarrow{AC}$
$= -(\mathbf{a} + \mathbf{b})$

(c) $\overrightarrow{AM} = \tfrac{1}{2}\overrightarrow{AC}$
$= \tfrac{1}{2}(\mathbf{a} + \mathbf{b})$

(d) In triangle MAB,

$\overrightarrow{MB} = \overrightarrow{MA} + \overrightarrow{AB} = -\overrightarrow{AM} + \overrightarrow{AB}$
$= -\tfrac{1}{2}(\mathbf{a} + \mathbf{b}) + \mathbf{a}$

$\therefore \qquad \overrightarrow{MB} = \tfrac{1}{2}(\mathbf{a} - \mathbf{b})$

3. PQRST is a polygon in which $\overrightarrow{PQ} = \mathbf{p}$, $\overrightarrow{QR} = \mathbf{q}$, $\overrightarrow{RS} = \mathbf{r}$, and $\overrightarrow{ST} = \mathbf{s}$.

(a) Find the vector represented by $\overrightarrow{PS}$.

(b) Describe the line that represents $\frac{1}{3}(\mathbf{p} + \mathbf{q} + \mathbf{r} + \mathbf{s})$.

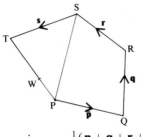

(a) From the diagram we see that $\overrightarrow{PS}$ is equivalent to $\overrightarrow{PQ} + \overrightarrow{QR} + \overrightarrow{RS}$

$$\therefore \qquad \overrightarrow{PS} = \mathbf{p} + \mathbf{q} + \mathbf{r}$$

(b) $\mathbf{p} + \mathbf{q} + \mathbf{r} + \mathbf{s}$ is represented by $\overrightarrow{PT}$

$\frac{1}{3}(\mathbf{p} + \mathbf{q} + \mathbf{r} + \mathbf{s})$ is represented by $\frac{1}{3}$ of $\overrightarrow{PT}$

$\therefore$ $\frac{1}{3}(\mathbf{p} + \mathbf{q} + \mathbf{r} + \mathbf{s})$ is represented by $\overrightarrow{PW}$ where $PW = \frac{1}{3}PT$.

4. A, P, Q and R are four points such that $\overrightarrow{AP} = 2\mathbf{a}$, $\overrightarrow{AQ} = 3\mathbf{b}$ and $\overrightarrow{AR} = 9\mathbf{b} - 4\mathbf{a}$. Show that P, Q and R are collinear.

Collinear means 'in the same straight line'. For P, Q and R to be collinear, PQ and QR (or PR) must be parallel.

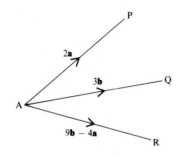

$$\overrightarrow{PQ} = \overrightarrow{PA} + \overrightarrow{AQ} = -2\mathbf{a} + 3\mathbf{b}$$

$$\overrightarrow{QR} = \overrightarrow{QA} + \overrightarrow{AR} = -3\mathbf{b} + (9\mathbf{b} - 4\mathbf{a})$$
$$= -4\mathbf{a} + 6\mathbf{b}$$

Hence $\qquad \overrightarrow{QR} = 2\overrightarrow{PQ}$

i.e. QR is parallel to PQ

$\therefore$ P, Q and R are collinear.

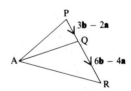

5. In the triangle OAB, $\overrightarrow{OA} = 3\mathbf{a}$, $\overrightarrow{OB} = 3\mathbf{b}$, $OQ = \frac{1}{3}OA$ and $OP = \frac{1}{3}OB$. AP and BQ meet at R.

(a) Express $\overrightarrow{PA}$ and $\overrightarrow{QB}$ each in terms of $\mathbf{a}$ and $\mathbf{b}$.

(b) Given that $QR = kQB$ and $PR = hPA$, find two expressions for $\overrightarrow{OR}$ in terms of $\mathbf{a}$, $\mathbf{b}$, h and k by considering

(i) $\triangle OQR$ (ii) $\triangle OPR$.

(c) By equating these two expressions find the value of k and hence find the ratio in which R divides QB.

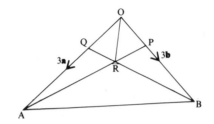

(a) $\overrightarrow{PA} = \overrightarrow{PO} + \overrightarrow{OA} = \frac{1}{3}\overrightarrow{BO} + \overrightarrow{OA} = -\mathbf{b} + 3\mathbf{a}$

$\overrightarrow{QB} = \overrightarrow{QO} + \overrightarrow{OB} = \frac{1}{3}\overrightarrow{AO} + \overrightarrow{OB} = -\mathbf{a} + 3\mathbf{b}$

(b) (i) $\overrightarrow{OR} = \overrightarrow{OQ} + \overrightarrow{QR} = \overrightarrow{OQ} + k\overrightarrow{QB}$

$\qquad\qquad\qquad = \mathbf{a} + k(3\mathbf{b} - \mathbf{a})$

(ii) $\overrightarrow{OR} = \overrightarrow{OP} + \overrightarrow{PR} = \mathbf{b} + h(3\mathbf{a} - \mathbf{b})$

(c) $\qquad\qquad \mathbf{a} + k(3\mathbf{b} - \mathbf{a}) = \mathbf{b} + h(3\mathbf{a} - \mathbf{b})$

$\Rightarrow \qquad (1 - k)\mathbf{a} + 3k\mathbf{b} = 3h\mathbf{a} + (1 - h)\mathbf{b}$

These vectors can be equal only if they are identical, i.e. the coefficients of both **a** and **b** must be equal.

i.e. $\qquad 1 - k = 3h \quad \text{and} \quad 3k = 1 - h$

$\therefore \qquad 1 - k = 3(1 - 3k) \qquad \Rightarrow \qquad 8k = 2$

Hence $\qquad k = \frac{1}{4}$

$\therefore \qquad QR = \frac{1}{4}QB \quad \text{and} \quad RB = \frac{3}{4}QB$

$\therefore \qquad$ R divides QB in the ratio $1:3$.

EXERCISE 3a HW · sewed

1.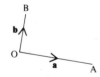

Express, in terms of **a** and **b**, the vector represented by
(a) $\overrightarrow{AB}$ (b) $\overrightarrow{BA}$.

2. Given the vectors **p**, **q** and **r** as shown, draw diagrams to illustrate

(a) $\mathbf{p} + \mathbf{r}$ (b) $\mathbf{q} - \mathbf{p}$ (c) $\mathbf{r} - \mathbf{q}$

(d) $\mathbf{p} + \mathbf{q} - \mathbf{r}$ (e) $\mathbf{r} - \mathbf{p} + \mathbf{q}$.

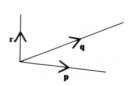

3. 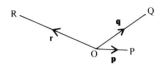 Find the vector represented by

(a) $\overrightarrow{PQ}$ (b) $\overrightarrow{QR}$ (c) $\overrightarrow{PR}$

Give your answers in terms of **p**, **q** and **r**.

4. In the diagram, $\overrightarrow{OA} = \mathbf{a}$, $\overrightarrow{OB} = \mathbf{b}$ and C is the midpoint of AB. Express each of the following vectors in terms of **a** and **b**.

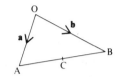

(a) $\overrightarrow{AB}$ (b) $\overrightarrow{AC}$ (c) $\overrightarrow{BC}$

5. In a quadrilateral ABCD, $\overrightarrow{AB} = \mathbf{a}$, $\overrightarrow{BC} = \mathbf{b}$, and $\overrightarrow{CD} = \mathbf{c}$. Express in terms of **a**, **b** and **c**,

(a) $\overrightarrow{AC}$ (b) $\overrightarrow{BD}$ (c) $\overrightarrow{DB}$ (d) $\overrightarrow{DA}$

6. Draw diagrams to illustrate each vector equation.

(a) $\overrightarrow{AB} = 2\overrightarrow{PQ}$ (b) $\overrightarrow{AB} - \overrightarrow{CB} = \overrightarrow{AC}$ (c) $\overrightarrow{AB} + \overrightarrow{BC} = 3\overrightarrow{AD}$

7. In triangle ABC, D bisects BC. Prove that $\overrightarrow{BA} + \overrightarrow{AC} = 2\overrightarrow{DC}$.

8. In a regular hexagon PQRSTU, $\overrightarrow{QR}$ represents a vector **a** and $\overrightarrow{UR}$ represents a vector **2b**.

Express in terms of **a** and **b** the vectors represented by

(a) $\overrightarrow{PQ}$ (b) $\overrightarrow{ST}$ (c) $\overrightarrow{QT}$

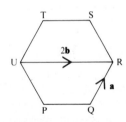

9. In triangle ABC, P and Q are the midpoints of AB and BC respectively. Use vectors to prove the midpoint theorem, i.e. that PQ is parallel to AC and half its length.

10. Given a pentagon ABCDE,

(a) express as a single vector

     (i) $\overrightarrow{AB} + \overrightarrow{BC} + \overrightarrow{CD}$

     (ii) $\overrightarrow{BC} + \overrightarrow{AB}$

     (iii) $\overrightarrow{AB} - \overrightarrow{AE}$

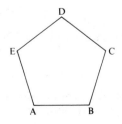

(b) find two ways in which $\overrightarrow{AD}$ can be expressed as a sum or difference of a number of vectors.

11. In the diagram, A, B and C are collinear. Find the value of k.

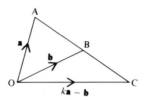

12. In the triangle ABC, $\overrightarrow{AB} = 2\mathbf{a}$, $\overrightarrow{AC} = 2\mathbf{b}$, E and F are the midpoints of AC and AB. BE and CF meet at G.

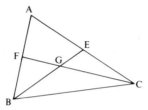

(a) Express $\overrightarrow{BE}$ and $\overrightarrow{CF}$ each in terms of $\mathbf{a}$ and $\mathbf{b}$.

(b) $EG = h\mathbf{EB}$ and $FG = k\mathbf{FC}$. By referring to triangles AFG and AEG, express $\overrightarrow{AG}$ as a sum of vectors in two different ways and hence find the values of h and k.

(c) In what ratio does G divide BE?

RESULTANTS AND COMPONENTS

At this stage in the book the work begins to require some knowledge of trigonometry. Any readers who have not yet acquired this skill should postpone going further with mechanics until they have studied the basics of trigonometry.

When two (or more) vectors are added, the single equivalent vector is called the resultant vector. The vectors that are combined are called *components*.

Consider the example of a heavy crate being pulled along by two ropes. The unit in which the forces in the ropes are measured is the newton (N).

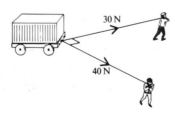

Although the ropes are pulling in different directions, the crate moves in only one direction. This is the direction of the resultant of the tensions (i.e. the pulling forces) in the ropes. By drawing a *triangle of vectors* we can find both the magnitude and the direction of the resultant force. (Remember that we use a larger arrow for the resultant.)

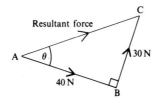

Note that a triangle of vectors *does not necessarily* give the *positions* of the components or the resultant. In this case, for example, each of the components acts on the crate so the equivalent resultant force also acts on the crate.

The magnitude and direction of the resultant force can be found by calculation.

In $\triangle$ABC:

$AC^2 = AB^2 + BC^2 \quad \Rightarrow \quad$ AC represents 50 N

and $\quad \tan A = \frac{3}{4} \quad (\tan = \frac{opp}{adj})$

$\Rightarrow \quad \hat{A} = 37°$ (nearest degree)

Hence the resultant force is of magnitude 50 N acting at an angle of 37° to AB. The calculation is easy because the components are at right angles.

This example can be extended to a general case where two vectors $\overrightarrow{AB}$ and $\overrightarrow{BC}$ are perpendicular. If the magnitudes of $\overrightarrow{AB}$ and $\overrightarrow{BC}$ are p and q, the magnitude of the resultant vector $\overrightarrow{AC}$ is given by $\sqrt{(p^2 + q^2)}$.

An alternative method is to draw, to scale, AB followed by BC and then join the *starting point to the finishing point*. The magnitude and direction of the resultant can then be measured from the drawing.

Scale drawing can be used also to find the resultant of any two or more vectors; this avoids the trigonometric calculations involved when the vectors are not perpendicular to each other.

Examples 3b

1. Find, by calculation, the resultant of two velocities if one is 7 km/h south west and the other is 12 km/h south east.

A sketch is drawn starting with a line representing one velocity followed by a line representing the other one.

AB represents 7 km/h south west.
BC represents 12 km/h south east.
AC represents the resultant velocity.

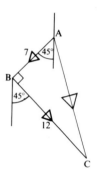

In $\triangle ABC$, $\quad AC^2 = AB^2 + BC^2 = 193$

$\therefore \qquad\qquad\qquad AC = 13.9 \,(3\,\text{sf})$

$\qquad\qquad \tan A = BC/AB = \frac{12}{7} = 1.714$

$\therefore \qquad\qquad\qquad A = 60° \,(\text{to the nearest degree})$

$\therefore \qquad$ the bearing of AC is $225° - 60° = 165°$

The resultant velocity is 13.9 km/h on a bearing of 165°.

2. A light aircraft is flying in still air at 180 km/h on a bearing of 052°. A steady wind suddenly springs up, blowing due south at 70 km/h. Find, by scale drawing, the velocity of the aircraft over the ground.

To find the resultant velocity we add the velocities of the plane and the wind.

Scale: 1 cm to 20 km/h

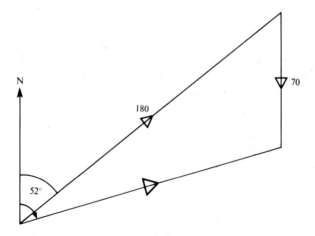

Measuring from the drawing gives the resultant velocity as

148 km/h on a bearing of 074°.

3. In a test of strength, four competitors each attempt to move a large sack of stones by pulling a rope attached to the sack. The magnitude of the force exerted by each competitor is shown in the diagram. Find the resultant force acting on the sack.

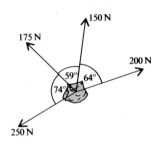

We can find the resultant by drawing lines representing, to scale, the forces taken in order and then joining the first point to the last.

Using a scale of 1 cm to 50 N, the forces are represented by AB, BC, CD and DE. The resultant is then represented by AE.

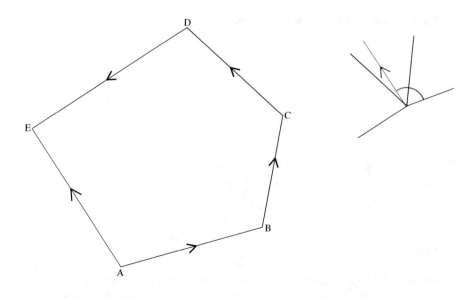

The resultant is a force of 219 N inclined at 108° to the force of 200 N.

If one of two components is given and the resultant is known, the other component can be found by adapting this method.

4. A boat leaves the harbour, A, and sails 3 km on a bearing of 040° to a position B. The boat's compass then goes wrong, so the boat heads for a buoy, C, which the skipper knows is 2 km north west of the harbour. How far, and on what bearing, did the boat travel during the second part of its trip?

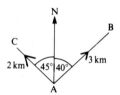

Let AB represent the first displacement, i.e. 3 km on 040°.
The second displacement is not known so it cannot yet be drawn.
The resultant displacement, 2 km north west, is represented by a line that starts from A and this line can be drawn; call it AC.

Now a scale drawing can be made.

Using 4 cm ≡ 1 km gives a drawing large enough for accurate measurements to be taken from, but, because space on the page is restricted, we have used 2 cm ≡ 1 km.

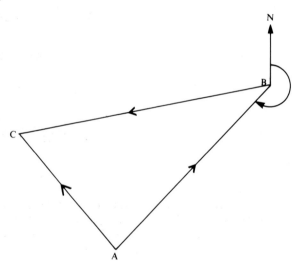

BC can now be drawn and measurements taken for distance and bearing.

The second part of the boat's trip is 3.5 km on a bearing of 255°.

Note that the direction of this journey must be from B to C, as C is the end point both for the components and for the resultant.

EXERCISE 3b

In questions 1 to 10, find the resultant of the given vectors. Use either calculation or scale drawing and remember always to give both the magnitude and the direction of the resultant.

1. A displacement of 12 km south followed by a displacement of 5 km east.

2. A displacement of 5 km east followed by a displacement of 12 km south. Is there any difference between the answer to this question and the answer to question 1?

3. A velocity of 24 m/s north and a velocity of 7 m/s east.

4. A force of 12 N west and a force of 16 N south.

5. Displacements of 10 m east and 12 m north east.

6. Two velocities, one of 4 m/s south and the other 5 m/s on a bearing 120°.

7. The two forces shown in the diagram (the direction of the resultant force can be given as an inclination to either of the given forces).

8. 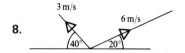 The two velocities shown in the diagram.

9. Displacements of 10 m east, 14 m north and 21 m on a bearing of 260°.

10. The forces shown in the diagram.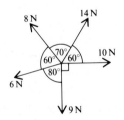

11. A man leaves his home by car and travels 5 km on a road running due east. The driver then turns left and travels 2 km due north to a junction where he joins a road that goes north west and drives a further 2 km. Find, from a scale drawing, his displacement from home by then.

12. An aircraft, flying with an engine speed of 400 km/h, is set on a course due north, in a wind of speed 60 km/h *from* the south west. At what speed and in what direction is the aircraft covering the ground?

13. On an orienteering exercise a woman starts from base and walks 500 m on a bearing of 138° and then 750 m on a bearing of 080°. What is her displacement from base? What bearing should she set to return to base?

In questions 14 and 15, a river running from north to south is flowing at 3 km/h.

14. A girl who can swim at 4 km/h is aiming directly across the river from east to west. Find the actual direction of her course across the river and the speed at which she passes over the river bed.

15. Another girl, who also swims at 4 km/h, notices that her friend is not moving straight across the river and works out how to go directly across. Draw a sketch of the way she does it and find the speed at which she crosses.

16. A plane leaves an airfield P, flying in a wind of 50 km/h blowing in a direction 048°. The plane arrives at an airfield Q, 400 km due east of P, 2 hours later. By making an accurate scale drawing find the plane's engine speed and the bearing the pilot set for the flight.

FINDING THE COMPONENTS OF A VECTOR

We have seen that two vectors can be combined into a single resultant vector. Now we will examine the reverse process, i.e. the replacing one vector by an equivalent pair of vectors. This process is called *resolving a vector into components*.

Suppose for example that a vector of magnitude 10 m east is to be replaced by two components, one of them due south and the other north east.

In the diagram, $\overrightarrow{AC}$ represents the given vector.
$\overrightarrow{AB}$ represents the component due south and
$\overrightarrow{BC}$ represents the component north east.
The directions of these components are known but their magnitudes are not.
However, the lengths of the lines AB and BC, and hence the magnitudes of the components, can be found.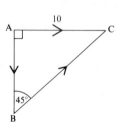

$\triangle ABC$ is an isosceles right-angled triangle, therefore $AB = AC = 10$.
Also Pythagoras' theorem gives $BC^2 = AB^2 + AC^2 \Rightarrow BC = 14.1$ (3 sf)

Therefore the required components are 10 m due south and 14.1 m north east.

Perpendicular Components

A vector can be resolved into an infinite variety of components in different directions but the most useful components, and the easiest to find, are a perpendicular pair.

Consider, for example, a plane taking off at an angle of 30° to the runway at 150 km/h.

The horizontal and vertical components of the velocity can be found by using trigonometry, i.e.

the horizontal component is given by $PQ = 150 \cos 30° = 130$ (3 sf)

and the vertical component is given by $QR = 150 \sin 30° = 75$.

Calculating the components of a vector plays a very important part in solving mechanics problems so it is important that they can be written down immediately in the form above.

Any reader who, up to now, would first have written down $\dfrac{PQ}{150} = \cos 30°$

should practice going straight to the form $PQ = 150 \cos 30°$; otherwise a great deal of time will be wasted.

Examples 3c

1. A skier ascends at a constant speed of 5 m/s in a chair lift inclined at 27° to the horizontal.

(a) Find the horizontal and vertical components of her velocity.

(b) What difference is there between the components found in part (a) and the horizontal and vertical components of the velocity of the chair as it returns to base at the same speed?

(a)

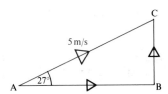

$\quad$ $AB = 5 \cos 27°$

∴ $\quad$ the horizontal component is 4.46 m/s (3 sf).

$\quad$ $BC = 5 \sin 27°$

∴ $\quad$ the vertical component is 2.27 m/s (3 sf).

(b) As the chair descends, each velocity component has the same magnitude, but is in the opposite direction, from that in part (a)
This can be indicated by a minus sign,

i.e. the horizontal component is −4.46 m/s
and the vertical component is −2.27 m/s.

2. A force of 98 N is pressing vertically downward on the inclined face of a wedge. If the angle of inclination of the wedge is 40° find the components of the force parallel to, and perpendicular to, the face of the wedge.

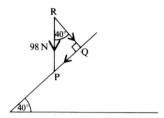

The required components, although not horizontal and vertical in this case, are perpendicular and can again be found from a right-angled triangle.

$$PQ = 98 \sin 40° = 62.99\ldots$$
$$QR = 98 \cos 40° = 75.07\ldots$$

Therefore, to 3 sf,   the component parallel to the face is 63.0 N
and the component perpendicular to the face is 75.1 N.

EXERCISE 3c

1. For each triangle write down an expression for the required side directly as a product of AB and a trig ratio of the given angle.

(a) 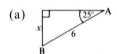   (c) 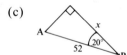   (e)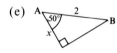

(b) 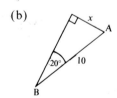   (d) 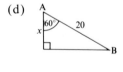   (f)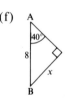

In each question from 2 to 7 find the horizontal and vertical components of the given vector, indicating the direction of each component by an arrow on a diagram.

2.   **3.**   **4.**

5.

6.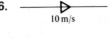

7.

In each question from 8 to 15 find the components parallel to and perpendicular to the inclined line.

8.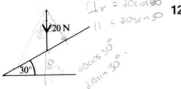

12.

9.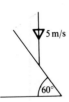

13.

10.

14.

11.

15.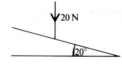

16. A stone is thrown up at an angle of 20° to the vertical with an initial velocity of 35 m/s. What are the initial horizontal and vertical components of the velocity of the stone?

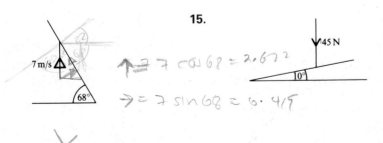

17. A train is travelling at 125 mph on a railway line that runs N 24°E and the direction of a canal is due east. Find the components, parallel and perpendicular to the canal, of the velocity of the train.

18. A boulder is falling vertically downward towards a hillside inclined at 23° to the horizontal. Its velocity just before impact is 176 m/s. Find the components of this velocity, parallel and perpendicular to the surface of the hillside.

19. The diagram shows a rectangular field with dimensions 120 m by 88 m. A boy pulls a truck directly from A to C with a force of 100 N. Find the components of this force parallel and perpendicular to AB.

CARTESIAN UNIT VECTOR NOTATION

A vector which has a magnitude of 1 is called a unit vector, irrespective of its direction.

There is, however, a set of special unit vectors each of which has a direction along one of the Cartesian axes of coordinates.

A unit vector in the direction of Ox is denoted by i,
a unit vector in the direction of Oy is denoted by j
and, for work in three dimensions,
a unit vector in the direction of Oz is denoted by k.

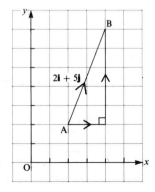

A vector $2\mathbf{i} + 5\mathbf{j}$ is made up of
 2 units in the positive direction of the x-axis
together with
 5 units in the positive direction of the y-axis,
i.e. $\overrightarrow{AB}$ represents the vector $2\mathbf{i} + 5\mathbf{j}$

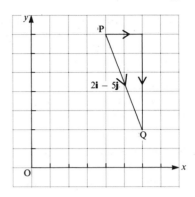

Similarly $2\mathbf{i} - 5\mathbf{j}$ is made up of
   2 units in the positive direction of Ox
together with
   5 units in the negative direction of Oy
i.e. $\overrightarrow{PQ}$ represents the vector $2\mathbf{i} - 5\mathbf{j}$

Any vector in the xy plane can be given in terms of a multiple of $\mathbf{i}$ together with a multiple of $\mathbf{j}$.

For this reason $\mathbf{i}$ and $\mathbf{j}$ are called *Cartesian base vectors*.

OPERATIONS ON CARTESIAN VECTORS

Modulus

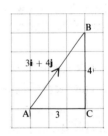

In this diagram $\overrightarrow{AB} = 3\mathbf{i} + 4\mathbf{j}$
The length of AB is $\sqrt{(3^2 + 4^2)} = 5$
This is the magnitude, or *modulus*, of $3\mathbf{i} + 4\mathbf{j}$
We denote the modulus of the vector by $|3\mathbf{i} + 4\mathbf{j}|$

i.e. $|3\mathbf{i} + 4\mathbf{j}| = 5$

Further, $\tan A = \frac{4}{3}$ $\Rightarrow$ $A = 53°$ (to the nearest degree)
so $3\mathbf{i} + 4\mathbf{j}$ is a vector with magnitude 5 units at $53°$ to AC.
This direction can also be described as being at $53°$ to the direction of $\mathbf{i}$.

Addition and Subtraction

We know that the resultant, i.e. the sum, of two vectors $\overrightarrow{AB}$ and $\overrightarrow{BC}$, is given by drawing AB followed by BC and joining A to C.

If $\overrightarrow{AB}$ and $\overrightarrow{BC}$ are given in $\mathbf{i}\,\mathbf{j}$ form, they can be drawn in the xy plane and their sum 'read' from the graph.

For example, if $\overrightarrow{AB} = 2\mathbf{i} + 5\mathbf{j}$
and $\overrightarrow{BC} = 7\mathbf{i} - 4\mathbf{j}$
$\overrightarrow{AB} + \overrightarrow{BC}$ is seen to be $9\mathbf{i} + \mathbf{j}$
which is

$$(2+7)\mathbf{i} + (5-4)\mathbf{j}$$

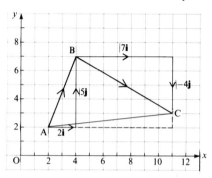

In general,

$$(a\mathbf{i} + b\mathbf{j}) + (p\mathbf{i} + q\mathbf{j}) = (a+p)\mathbf{i} + (b+q)\mathbf{j}$$

i.e. we add the coefficients of $\mathbf{i}$ and add the coefficients of $\mathbf{j}$.

Now suppose that we want to *subtract* $7\mathbf{i} - 4\mathbf{j}$ from $2\mathbf{i} + 5\mathbf{j}$.

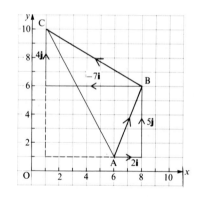

We draw $2\mathbf{i} + 5\mathbf{j}$ followed by a vector
equal to $7\mathbf{i} - 4\mathbf{j}$ but in the opposite
direction, i.e. $-7\mathbf{i} + 4\mathbf{j}$

The resultant this time is seen to
be $-5\mathbf{i} + 9\mathbf{j}$ which is

$$(2-7)\mathbf{i} + (5-[-4])\mathbf{j}$$

In general

$$(a\mathbf{i} + b\mathbf{j}) - (p\mathbf{i} + q\mathbf{j}) = (a-p)\mathbf{i} + (b-q)\mathbf{j}$$

This time we subtract the coefficients of $\mathbf{i}$ and of $\mathbf{j}$.

Three-dimensional vectors given in $\mathbf{i}\,\mathbf{j}\,\mathbf{k}$ form can be added or subtracted in the
same way, i.e.

$$(a\mathbf{i} + b\mathbf{j} + c\mathbf{k}) \pm (p\mathbf{i} + q\mathbf{j} + r\mathbf{k}) = (a\pm p)\mathbf{i} + (b\pm q)\mathbf{j} + (c\pm r)\mathbf{k}$$

For example

$$(2\mathbf{i} + 3\mathbf{j} - 5\mathbf{k}) + (\mathbf{i} - 7\mathbf{j} + 2\mathbf{k}) = \mathbf{i} - 4\mathbf{j} - 3\mathbf{k}$$

and $\quad (8\mathbf{i} - \mathbf{j} + \mathbf{k}) - (2\mathbf{i} - 5\mathbf{j} + 3\mathbf{k}) = 6\mathbf{i} + 4\mathbf{j} - 2\mathbf{k}$

Examples 3d

1. Taking **i** as a unit vector due east and **j** as a unit vector due north, express in the form $a\mathbf{i} + b\mathbf{j}$ a vector of magnitude 24 units on a bearing of 220°.

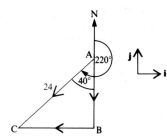

$$AB = 24 \cos 40° = 18.4 \text{ (3 sf)}$$
$$BC = 24 \sin 40° = 15.4 \text{ (3 sf)}$$
$$\overrightarrow{AC} = \overrightarrow{AB} + \overrightarrow{BC}$$
$$= -18.4\mathbf{j} + (-15.4\mathbf{i})$$

The required vector is $-15.4\mathbf{i} - 18.4\mathbf{j}$

2. A vector **V** is in the direction of the vector $12\mathbf{i} - 5\mathbf{j}$ and its magnitude is 39. Find **V** in the form $a\mathbf{i} + b\mathbf{j}$.

V is parallel to $12\mathbf{i} - 5\mathbf{j}$

$$\therefore \qquad \mathbf{V} = k(12\mathbf{i} - 5\mathbf{j}) = 12k\mathbf{i} - 5k\mathbf{j}$$

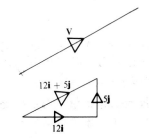

$$|\mathbf{V}| = |12k\mathbf{i} - 5k\mathbf{j}|$$
$$= \sqrt{[(12k)^2 + (-5k)^2]}$$
$$= \sqrt{(169k^2)} = 13k$$

But $\quad |\mathbf{V}| = 39$

$$\therefore \qquad 13k = 39 \quad \Rightarrow \quad k = 3$$
$$\therefore \qquad \mathbf{V} = 36\mathbf{i} - 15\mathbf{j}$$

EXERCISE 3d

1. Express each given vector in the form $a\mathbf{i} + b\mathbf{j}$

(a)

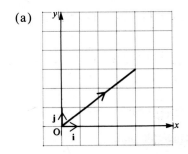

(b)

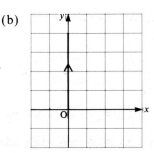

(c)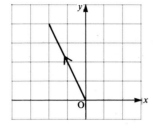

(e)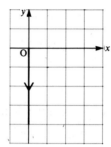

(d)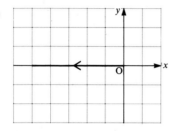

(f)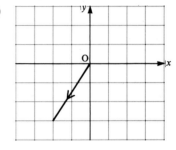

2. Find the modulus of each vector given in question 1.

3. Taking **i** as a unit vector due east and **j** as a unit vector due north, express in the form $a\mathbf{i} + b\mathbf{j}$,

    (a) a vector of magnitude 14 units on a bearing 060°

    (b) a vector of magnitude 20 units on a bearing 180°

    (c) a velocity of 75 units south west

    (d) a displacement of 400 units in the direction of $3\mathbf{i} + 4\mathbf{j}$.

In each question from 4 to 6, find (a) $\mathbf{a} + \mathbf{b}$ (b) $\mathbf{a} - \mathbf{b}$
and illustrate the results in a diagram.

4. $\mathbf{a} = 6\mathbf{i} + \mathbf{j},\quad \mathbf{b} = 3\mathbf{i} - 5\mathbf{j}$

5. $\mathbf{a} = \mathbf{i},\quad \mathbf{b} = -\mathbf{j}$

6. $\mathbf{a} = -2\mathbf{i} - \mathbf{j},\quad \mathbf{b} = \mathbf{i} - \mathbf{j}$

In questions 7 to 11, $\mathbf{p} = 4\mathbf{i} - 3\mathbf{j},\quad \mathbf{q} = -12\mathbf{i} + 5\mathbf{j},\quad \mathbf{r} = \mathbf{i} - \mathbf{j}$

7. Find (a) $|\mathbf{p}|$ (b) $|\mathbf{q}|$ (c) $|\mathbf{r}|$, giving the answer as a square root.

8. Find a vector in the direction of **p** and with magnitude 30 units.

9. Find (a) the resultant of **q** and **r** (b) $|\mathbf{q}+\mathbf{r}|$

10. Find two vectors, each with a magnitude twice that of **p** and parallel to **p** (remember that parallel vectors can be in the same or opposite directions).

11. Find a vector **v** such that

    (a) $|\mathbf{v}| = 35$ and **v** is in the direction of $\mathbf{q}+5\mathbf{r}$

    (b) **v** is parallel to $\mathbf{p}-\mathbf{q}$ and is half the size of $\mathbf{p}-\mathbf{q}$.

CHAPTER 4

FORCE

THE CONCEPT OF FORCE

So far in this book we have been dealing mainly with motion (kinematics). Now we must consider how motion of different kinds is caused (dynamics).

Force is such an everyday quantity that we all have an intuitive idea of what it is. We know that to move a heavy cabinet across a room we push it; to raise a bucket of cement from ground level to roof height, a builder pulls it up. Pushes and pulls are both forces and these simple situations illustrate that force is needed to make an object start to move.

On the other hand, a runaway shopping trolley can be stopped either by getting in front of it and exerting a push, or by holding it from behind and pulling. So a force can also cause a moving object to begin to stop.

Now consider what happens when someone who is holding a stone lets it drop. The stone begins to move downwards. What makes it move? Something must be *pulling it* down. That something is the weight of the stone; *weight is the name we give to the force that attracts each object to the earth, i.e. the force due to gravity.*

A book resting on a horizontal surface, on the other hand, does not fall down so a force must be preventing it from falling. This is the force exerted by the surface to hold the book up. As the book does not move we deduce that this force (upward) and the weight of the book (downward) must balance.

When an object does not move, we say that the object, and the forces that act on it, are *in equilibrium.*

Conventions

Mechanics is a complex subject and it is impossible to deal with all aspects of the topic at the beginning. So certain simplifications have to be made in order to deal with one idea at a time. You will see that, although some of these conventions are too ideal to be factual, they are reasonably close approximations to reality and make it possible for a student to absorb the principles of mechanics without being hampered by too many details at this stage.

A small object is regarded as existing at a point and is called a *particle*.

An object whose mass is small is considered to be weightless. It is called *light* and its mass is ignored. Such an object may be a particle or it can be a fine string or wire.

A flat shape whose thickness is small is regarded as being two-dimensional and is called a *lamina*.

Note that 'small' is a comparative term, e.g. a person could be thought of as a 'small object' and treated as a weightless particle in relation to the Forth Bridge, but certainly not in relation to a chair.

Air resistance is usually ignored (on the grounds that it is generally small compared with other forces).

Any object attached rigidly to the earth is called *fixed* and is considered to be immoveable. Forces that act on a fixed object are disregarded.

TYPES OF FORCE

Forces of Attraction

Gravitational attraction is the most important force of this type and almost the only one we shall meet in this book. The gravitational attraction of the earth on an object (also referred to as a body) is called the weight of the object and its acts vertically downward on the object. It is almost always given the symbol W.

The effect of weight can be seen when an object falls and also it can be felt when an object is held, i.e. *the weight of an object acts on it at all times whether the object is moving or not.*

(There are other forces of attraction such as those between a magnet and an iron object.)

Contact Forces

Consider again a book resting in contact with a horizontal surface. The force exerted upward on the book by the surface is a *contact force* called a *normal reaction*. A normal reaction acts in a direction that is perpendicular to the surface of contact and away from that surface. So in this case the *normal reaction* acts vertically upward on the book.

The two forces acting on the book, i.e. the normal reaction, R, and the weight, W, are shown in the diagram.

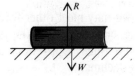

Now suppose that a small push *P* is applied horizontally to the book. If the book and the surface do not have a slippery covering, it is quite likely that the book will not move. Why not? There must be another force, equal and opposite to the push, balancing it. This is a frictional force. Friction can occur only when objects are in rough contact, so it is another example of a contact force.

A frictional force acts on a body *along the surface of contact* and in a direction which *opposes the potential movement* of that body.

This diagram shows all the forces acting on the book.

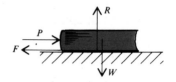

It is very rare for there to be no friction at all between an object and a surface but there can be so little that its effect can be discounted. In this case the contact is said to be *smooth*.

Forces of Attachment

Consider a mass, hanging by a string from a fixed point. The mass does not move so its weight acting downward must be balanced by an upward force.
This force is the *tension* in the string; it is a force of attachment.

Note that a string can never push and it can pull only if it is taut. For a simple attachment a string cannot be taut at one end and slack at the other, so a taut string exerts an *inward* pull at each end on the object which is attached at that end.

Another way in which objects can be attached to each other is by means of a hinge or pivot. So the force exerted by a hinge is another force of attachment. One difference between it and the tension in a string is that a hinge *can* push as well as pull; another is that the direction of a hinge force is not usually known whereas *the tension in a string acts along the string*.

We may meet other forces occasionally, such as wind force, the driving force of an engine etc., but the vast majority of the forces involved at this level belong to the three types described above, i.e. gravitational attraction, contact and attachment.

DRAWING DIAGRAMS

When considering any situation concerning the action of forces on a body the first, and vital, step is to draw a clear, uncomplicated diagram of the forces acting on the object.
Some useful points to remember are:

Unless the object is light, its weight acts vertically downwards.

If the body is in contact with another surface, a normal reaction always acts on the body. In addition, unless the contact is smooth, there may be a frictional force.

If the body is attached to another by a string or hinge, a force acts on the body at the point of attachment.

There is a common misconception that, all the time an object is moving, there *has* to be a force in the direction of motion. This is not true. One of the types of force described earlier in the chapter *may* be acting in the direction of motion but if none of them is, do not fall into the trap of introducing an 'extra' force.

Two or more objects may be linked, or may be in contact with each other. When considering *one* of these objects, make sure that you draw only those forces that act on that separate part and do not include forces which apply to another part of the system. This applies particularly in problems where two objects are attached to each other; if they are connected by a string, one object is affected only by the tension at the end of a string to which it is attached – the tension at the other end acts on the other object.

Finally, do not draw too small a diagram and make the force lines long enough to be seen clearly.

Examples 4a

Each of the following examples describes a situation and shows how a working diagram can be drawn.

1. A small block is sliding down a smooth inclined plane.

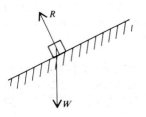

The normal reaction R is perpendicular to the plane which is the surface of contact.

2. A load is being pulled along a rough horizontal floor by a rope inclined at 50°
to the floor.

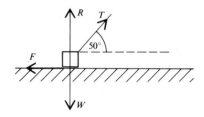

The frictional force acts along the plane in the direction opposite to the motion of the load.

3. A ladder rests with its foot on rough ground and the top against a smooth
wall.

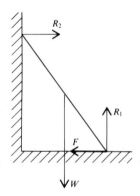

At the foot, the normal reaction is perpendicular to the ground. The frictional force acts along the
ground and towards the wall because, *if* the ladder moved, its foot would slip *away* from the wall.

At the top of the ladder the normal reaction is perpendicular to the wall and there is no friction. The
weight of the ladder acts through the centre of gravity (the point of balance of the ladder).

4. A beam is hinged at one end to a wall to which it is inclined at 60°. It is held
in this position by a horizontal chain attached to the other end.

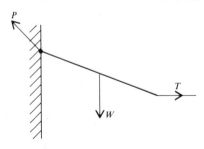

The direction of the force that the hinge exerts on the beam is not known at this stage so we cannot
mark it at any specific angle.
(Later on we will see how to determine this angle.)

5. A truck is attached by a rope to an electric-powered engine which is being driven along a horizontal smooth track.

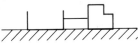

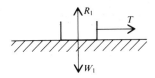

Considering the truck, the only forces acting are the weight of the truck, the vertical normal reaction and the tension in the rope which acts away from the truck (i.e. inward along the rope). Note that the driving force of the engine *does not* act on the truck, it is the tension in the rope that pulls the truck forward.

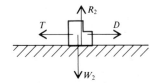

Acting on the engine alone we have the weight, the normal reaction, the driving force and the tension in the rope which acts towards the centre of the rope (i.e. it is a drag on the engine).

6. A particle is fastened to one end of a light string. The other end of the string is held and the particle is whirled round in a circle in a vertical plane.

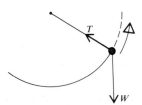

The only forces acting on the particle are its weight and the tension in the string. There is *no force in the direction of motion of the particle.*

EXERCISE 4a

In each question from 1 to 6, copy the diagram and draw the forces acting on the given object in the specified situation.

1.

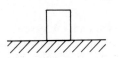

A block at rest on a smooth horizontal surface.

2.

A plank resting on two supports, one at each end.

3.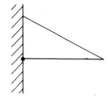

A rod hinged to a wall and held in a horizontal position by a string.

5.

A small block at rest on a rough inclined plane.

4.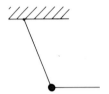

A light string fixed at one end has a particle tied to the other end being pulled aside by a horizontal force.

6.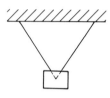

A small picture supported by two cords.

For the remaining questions draw a diagram of the specified object and mark on the diagram all the forces acting on the object.

7. A ladder with its foot on rough ground is leaning against a rough wall.

8. A particle is attached to one end of a light string whose other end is fixed to a point A. The particle is

    (a)   hanging at rest

    (b)   rotating in a horizontal circle below A

    (c)   held at 30° to the vertical by a horizontal force.

9. A stone that has been thrown vertically up into the air when it is

    (a)   going up   (b)   coming down.

10. A rod of length 1 m is hinged at one end to a wall. It is held in a horizontal position by a string joining the point of the rod that is 0.8 m from the wall, to a point on the wall 1 m vertically above A.

11. A shelf AB is supported by two vertical strings, one at each end. A vase is placed on the shelf, a quarter of the way along the shelf from A. Draw separate diagrams to show the forces acting on

    (a)   the shelf   (b)   the vase.

12. Two bricks are placed, one on top of the other, on a horizontal surface. Draw separate diagrams to show the forces acting on

    (a) the top brick

    (b) the lower brick.

13. A beam is hinged at one end A to a wall and is held horizontal by a rope attached to the other end B and to a point on the wall above A. The rope is at 45° to the wall. A crate hangs from B. Draw separate diagrams to show the forces acting on

    (a) the beam

    (b) the crate.

14. The diagram shows a rough plank resting on a cylinder and with one end of the plank on rough ground.

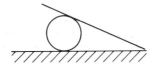

    Draw diagrams to show:

    (a) the forces acting on the plank,

    (b) the forces acting on the cylinder.

15. A person standing on the edge of a flat roof lowers a package over the edge, by a rope, for a colleague to collect. Draw diagrams to show

    (a) the forces acting on the package

    (b) the forces acting on the person on the roof.

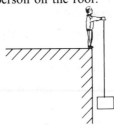

FINDING THE RESULTANT OF COPLANAR FORCES

It is quite common, as we have seen from the examples above, for an object to be under the action of several coplanar forces (i.e. all in one plane) in different directions. To investigate the overall effect of these forces we need to be able to find their resultant.

In Chapter 3 we saw that the magnitude and direction of the resultant of two perpendicular forces was easy to find. For more than two forces, however, scale drawing and measurement was used. As this method gives only a rough result we are now going to consider a way to *calculate* the resultant of any set of coplanar forces.

To do this we choose two perpendicular directions and find the components of all the forces in these directions; components in a chosen direction are positive while those in the opposite direction are negative.
By collecting each set of components we can now replace the original set of forces by an equivalent pair of forces in perpendicular directions.

Note that defining a direction as, say, 'along an inclined plane' is ambiguous because it could mean either up or down the plane. A simple way to clarify the definition is to add an arrow in the correct direction.

Consider, for example, a small block resting on a rough plane inclined to the horizontal at 30°. The forces acting on the block are shown in the diagram.

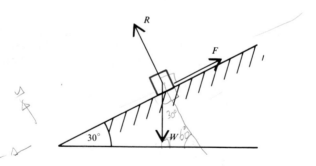

As the normal reaction and the frictional force are perpendicular to each other, it is sensible to find the components of each force in these two directions, i.e. along (↗) and perpendicular (↖) to the plane, i.e.

	Friction	Reaction	Weight
Component ↗	$-F$	0	$W \sin 30°$
Component ↖	0	R	$-W \cos 30°$

Now we can collect the components of force down and perpendicular to the plane and to indicate these operations we write:

Resolving ↗   gives $W \sin 30° - F$ [1]

Resolving ↖   gives $R - W \cos 30°$ [2]

Calculating the Resultant

If the expression [1] above is represented by X and expression [2] by Y, we have

$$X = W \sin 30° - F$$
$$Y = R - W \cos 30°$$

The magnitude of the resultant, R, of X and Y is

$$\sqrt{(X^2 + Y^2)}$$

and R makes an angle α with the plane where

$$\tan \alpha = Y/X$$

In a case where each of the forces is given in the form $a\mathbf{i} + b\mathbf{j}$ the forces are already expressed as components in the directions of $\mathbf{i}$ and $\mathbf{j}$ and it remains only to find X, Y and R.

Examples 4b

1. A ladder of weight W rests with its top against a rough wall and its foot on rough ground which slopes down from the base of the wall at 10° to the horizontal. Resolve, horizontally and vertically, all the forces acting on the ladder.

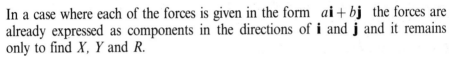

Drawing the components of R_2 and F_2 on separate small diagrams can help.

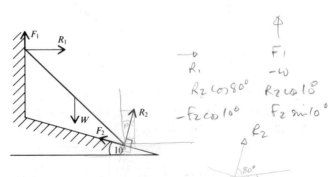

Resolving → gives $\quad R_1 - F_2 \cos 10° + R_2 \sin 10°$

Resolving ↑ gives $\quad F_1 - W + F_2 \sin 10° + R_2 \cos 10°$

2. Find the magnitude of the resultant of the set of forces $3\mathbf{i} + 5\mathbf{j}$, $-7\mathbf{j}$, $-4\mathbf{i} + 11\mathbf{j}$, $5\mathbf{i}$ and $\mathbf{i} + 3\mathbf{j}$ Each force is measured in newtons. Find the angle between the resultant and the unit vector $\mathbf{i}$.

Let the resultant be $X\mathbf{i} + Y\mathbf{j}$

$$X\mathbf{i} + Y\mathbf{j} = (3\mathbf{i} + 5\mathbf{j}) + (-7\mathbf{j}) + (-4\mathbf{i} + 11\mathbf{j}) + 5\mathbf{i} + (\mathbf{i} + 3\mathbf{j})$$
$$= (3 - 4 + 5 + 1)\mathbf{i} + (5 - 7 + 11 + 3)\mathbf{j}$$
$$= 5\mathbf{i} + 12\mathbf{j}$$

$$|5\mathbf{i} + 12\mathbf{j}| = \sqrt{(25 + 144)} = 13$$
$$\tan\alpha = \tfrac{12}{5} = 2.4$$
$$\alpha = 67° \text{ (nearest degree)}$$

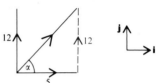

The resultant is 13 N at 67° to $\mathbf{i}$.

3. Find the resultant of the forces of 4, 6, 2 and 3 newtons shown in the diagram.

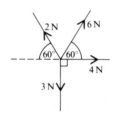

Let the resultant have components X and Y newtons in the directions shown.

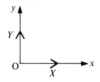

Resolving the forces along Ox and Oy we have:

Resolving $\longrightarrow$ $X = 4 + 6\cos 60° - 2\cos 60°$ (remember that $\cos 60° = \tfrac{1}{2}$)
$$= 4 + 3 - 1 = 6$$

Resolving $\uparrow$ $Y = 6\sin 60° - 3 + 2\sin 60°$
$$= 8 \times 0.8660 - 3 = 3.928$$

If the resultant force is R newtons

$$R = \sqrt{(X^2 + Y^2)} = \sqrt{(6^2 + 3.928^2)}$$
$$= 7.17 \quad (3\text{ sf})$$

and $\tan\alpha = \dfrac{Y}{X} = \dfrac{3.928}{6} = 0.6546\ldots$

$\Rightarrow$ $\alpha = 33°$ (nearest degree)

Therefore the resultant force is 7.17 N at 33° to the force of 4 N.

Sometimes the directions of a group of forces are given with reference to the sides, diagonals etc. of a polygon; the magnitudes of the forces are given separately.

It is important to realise that, in such cases, the forces are *not necessarily* represented by the *lengths* or *positions* of the lines in the polygon.

4. ABCDEF is a regular hexagon. Four forces act on a particle. The forces are of magnitudes 3 N, 4 N, 2 N and 6 N and they act in the directions of the sides AB, AC, EA and AF respectively.
   Find the magnitude of the resultant force and the angle it makes with AB.

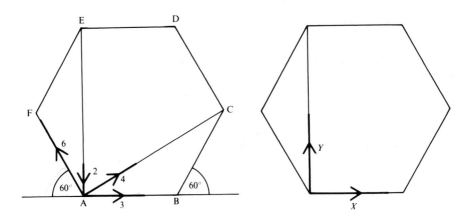

Let the resultant have components X newtons and Y newtons, parallel and perpendicular to AB as shown.

Resolving   → $X = 3 + 4 \cos 30° - 6 \cos 60°$

$$= 3 + 3.464 - 3 = 3.464$$

Resolving   ↑ $Y = 4 \sin 30° - 2 + 6 \sin 60°$

$$= 2 - 2 + 5.196 = 5.196$$

If R newtons is the resultant force then, correct to 3 sf,

$$R^2 = X^2 + Y^2 = 12 + 27 = 39$$

∴ $R = \sqrt{39}$

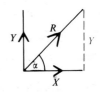

and $\tan \alpha = \dfrac{Y}{X} = \dfrac{5.196}{3.464} = 1.5$

The resultant force is $\sqrt{39}$ N at 56° to AB (nearest degree).

EXERCISE 4b

In each question from 1 to 6, find the magnitude of the resultant of the given vectors and give the angle between the resultant and the direction of the positive *x*-axis.

1.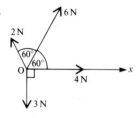

4.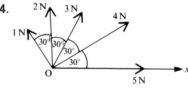

2.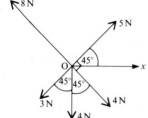

5.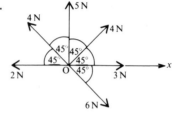

3.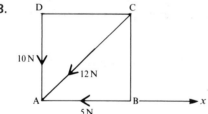

ABCD is a square.

6.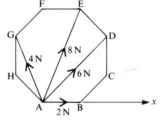

ABCDEFGH is a regular octagon.

In each question from 7 to 10,

(a) illustrate the vectors by a sketch

(b) express, in the form $a\mathbf{i} + b\mathbf{j}$,   the resultant of the given vectors

(c) find the magnitude of the resultant and give the angle that the resultant makes with the vector **i**.

7. Four forces, measured in newtons, represented by

$$4\mathbf{i} - 3\mathbf{j}, \quad \mathbf{i} + 6\mathbf{j}, \quad -2\mathbf{i} + 5\mathbf{j}, \quad \text{and} \quad 3\mathbf{i}$$

8. Velocities, in metres per second, represented by

$$4\mathbf{i} - 7\mathbf{j}, \quad -3\mathbf{i} + 8\mathbf{j}, \quad 2\mathbf{i} + 3\mathbf{j}, \quad 8\mathbf{i} \quad \text{and} \quad \mathbf{i} + \mathbf{j}$$

9. Displacements, measured in metres, represented by

$$-6\mathbf{i} + \mathbf{j}, \quad 2\mathbf{i} - 5\mathbf{j}, \quad \mathbf{i} + 4\mathbf{j} \quad \text{and} \quad 3\mathbf{i} + 2\mathbf{j}$$

10. Forces, in newtons, represented by

$$2\mathbf{i} + 2\mathbf{j}, \quad \mathbf{i} - 7\mathbf{j}, \quad -6\mathbf{i} + \mathbf{j}$$

11. ABCD is a rectangle in which $AB = 4$ m and $BC = 3$ m. A force of magnitude 3 N acts along AB towards B. Another force of magnitude 4 N acts along AC towards C and a third force, 3 N, acts along AD towards D. Find the magnitude of the resultant of these forces and find the angle the resultant makes with AD.

12. A surveyor starts from a point O and walks 200 m due north. He then turns clockwise through $120°$ and walks 100 m after which he walks 300 m due west. What is his resultant displacement from O?

13. Three boys are pulling a heavy trolley by means of three ropes. The boy in the middle is exerting a pull of 100 N. The other two boys, whose ropes both make an angle of $30°$ with the centre rope, are pulling with forces of 80 N and 140 N. What is the resultant pull on the trolley and at what angle is it inclined to the centre rope?

14. Starting from O, a point P traces out consecutive displacement vectors of
$2\mathbf{i} + 3\mathbf{j}, \quad -\mathbf{i} + 4\mathbf{j}, \quad 7\mathbf{i} - 5\mathbf{j} \quad \text{and} \quad \mathbf{i} + 3\mathbf{j}.$
What is the final displacement of P from O?

15. A river is flowing due east at a speed of 3 m/s. A boy in a rowing boat, who can row at 5 m/s in still water, starts from a point O on the south bank and steers the boat at right angles to the bank. The boat is also being blown by the wind at 4 m/s south-west. Taking axes Ox and Oy in the directions east and north respectively find the velocity of the boat in the form $p\mathbf{i} + q\mathbf{j}$ and hence find its resultant speed.

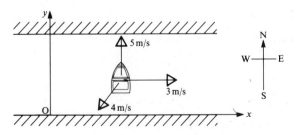

16. A small boat is travelling through the water with an engine speed of 8 km/h. It is being steered due east but there is a current running south at 2 km/h and wind is blowing the boat south-west at 4 km/h. Find the resultant velocity of the boat.

17. Velocities of magnitudes 5 m/s, 7 m/s, 4 m/s and 6 m/s act in the directions north-east, north, south-east and west respectively. Taking **i** and **j** as unit vectors east and north respectively,

    (a) draw a sketch showing the separate velocities

    (b) find, in the form $a\mathbf{i} + b\mathbf{j}$, the resultant velocity

    (c) find the bearing of the resultant velocity

    (d) find the resultant speed.

18. ABC is an equilateral triangle and D is the midpoint of BC. Forces of magnitudes 8 N, 6 N and 12 N act along AB, AC and DA respectively (the order of the letters gives the direction of the force). Find the magnitude of the resultant force and the angle between the resultant and DA.

CONSOLIDATION A

SUMMARY

Motion

For motion with constant speed:

$$\text{distance} \ = \ \text{speed} \times \text{time}$$

$$\text{average speed} \ = \ \frac{\text{total distance}}{\text{total time}}$$

Velocity is the rate of increase of displacement.

Acceleration is the rate of increase of velocity.

Using a displacement–time graph

The velocity at a particular time, t, is given by the gradient of the tangent to the graph at that value of t.

Using a velocity–time graph

The acceleration at a particular time, t, is given by the gradient of the tangent to the graph at that value of t.

The displacement after time t is given by the area under the graph for that time interval.

For motion with constant acceleration

Using the notation

$$u = \text{initial velocity} \quad v = \text{final velocity} \quad a = \text{acceleration}$$
$$t = \text{time interval} \quad \text{and} \quad s = \text{displacement}$$

the equations of motion are:

$$v \ = \ u + at$$
$$s \ = \ \tfrac{1}{2}(u + v)t$$
$$s \ = \ ut + \tfrac{1}{2}at^2 \quad \text{and} \quad s \ = \ vt - \tfrac{1}{2}at^2$$
$$v^2 - u^2 \ = \ 2as$$

Vectors

A quantity that has both magnitude and direction is a vector.
A vector can be represented by a line segment.

The *magnitude* of the vector is represented by the *length of the line* and the direction of the vector by the direction of the line.

Two *parallel vectors* with the same magnitude are:
   equal if they are in the same direction
   equal and opposite if they are in opposite directions.

If two vectors **a** and **b** are parallel then $\mathbf{a} = k\mathbf{b}$.

When, starting from a point A, lines representing vectors in magnitude and direction are drawn consecutively, the line starting at A that completes the polygon represents the *resultant* vector; all the other vectors are *components*.

The resultant of two perpendicular vectors, **p** *and* **q**:
   is of magnitude $\sqrt{(p^2 + q^2)}$
   makes an angle α with the vector **p** where $\tan \alpha = p/q$

The resultant of a set of coplanar vectors can be *calculated* by resolving all forces in two perpendicular directions.

Vectors of unit magnitude in the directions of Ox and Oy are denoted respectively by **i** and **j**.

A vector in the x–y plane whose components in the directions Ox and Oy are a and b respectively, can be written $a\mathbf{i} + b\mathbf{j}$.

The magnitude of the vector $a\mathbf{i} + b\mathbf{j}$ is denoted by $|a\mathbf{i} + b\mathbf{j}|$ and is of value $\sqrt{(a^2 + b^2)}$.

Types of Force

Contact forces occur when solid objects are in contact. A pair of equal and opposite forces act, one on each of the objects, perpendicular to the surface of contact.

Forces of attachment act when two objects are connected by, for example, a string or a hinge. Two equal and opposite forces act, one on each of the objects; in the case of a string the forces are an inward pull at each end of the string; the directions of the forces at a hinge are, in general, not known.

The earth exerts a force of attraction on any body outside its surface. This force is the weight of the object. The acceleration produced by the weight of an object is denoted by g; the approximate value of this acceleration at the surface of the earth is 9.8 m/s^2.

MISCELLANEOUS EXERCISE A

Each question from 1 to 5 is followed by several suggested responses. Choose the correct response.

1. The area shaded in the graph represents

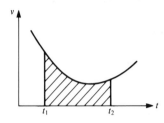

  A the average velocity in the time interval $(t_2 - t_1)$.

  B the increase in the displacement in the time interval $(t_2 - t_1)$.

  C the acceleration at time t_2.

2. If ABCD is a quadrilateral whose sides represent vectors, $\overrightarrow{AB}$ is equivalent to

  A $\overrightarrow{CA} + \overrightarrow{CB}$ **B** $\overrightarrow{CD}$ **C** $\overrightarrow{AC} - \overrightarrow{BC}$

3. $\overrightarrow{AB}$ and $\overrightarrow{PQ}$ are two vectors such that $\overrightarrow{AB} = 2\overrightarrow{PQ}$.

  A AB is parallel to PQ.

  B PQ is twice as long as AB.

  C A, B, P and Q must be collinear.

4. Two forces $\mathbf{F}_1$ and $\mathbf{F}_2$ have a resultant $\mathbf{F}_3$. If $\mathbf{F}_1 = 2\mathbf{i} - 3\mathbf{j}$ and $\mathbf{F}_3 = 5\mathbf{i} + 4\mathbf{j}$ then $\mathbf{F}_2$ is

  A $7\mathbf{i} + \mathbf{j}$ **B** $-3\mathbf{i} - 7\mathbf{j}$ **C** $3\mathbf{i} + 7\mathbf{j}$

5. When a number of particles, all of different weights, are dropped, the acceleration of each particle

  A is constant but different for each particle, depending on its weight.

  B is constant and the same for each particle.

  C increases as the particle falls.

6. Two forces $(3\mathbf{i} + 2\mathbf{j})$ N and $(-5\mathbf{i} + \mathbf{j})$ N act at a point. Find the magnitude of the resultant of these forces and determine the angle which the resultant makes with the unit vector $\mathbf{i}$. (AEB)

7. O is the origin, and points A and B have position vectors relative to O given by

$$\overrightarrow{OA} = \mathbf{i} + 7\mathbf{j} \quad \text{and} \quad \overrightarrow{OB} = 5\mathbf{i} + 5\mathbf{j}.$$

Show that the vectors $\overrightarrow{OA}$ and $\overrightarrow{OB}$ have equal magnitudes.

Points C and D have position vectors given by

$$\overrightarrow{OC} = 2\overrightarrow{OA} \quad \text{and} \quad \overrightarrow{OD} = \overrightarrow{OA} + \overrightarrow{OB}.$$

Express $\overrightarrow{OC}$ and $\overrightarrow{OD}$ in terms of $\mathbf{i}$ and $\mathbf{j}$, and draw a diagram showing the positions of A, B, C and D. (UCLES)

8. Three forces $(3\mathbf{i} + 5\mathbf{j})$ N, $(4\mathbf{i} + 11\mathbf{j})$ N, $(2\mathbf{i} + \mathbf{j})$ N act at a point. Given that $\mathbf{i}$ and $\mathbf{j}$ are perpendicular unit vectors find

(a) the resultant of the forces in the form $a\mathbf{i} + b\mathbf{j}$

(b) the magnitude of this resultant

(c) the cosine of the angle that the resultant makes with the unit vector $\mathbf{i}$.

(AEB)

9. The diagonals of the plane quadrilateral ABCD intersect at O, and X, Y are the midpoints of the diagonals AC, BD respectively. Show that

(a) $\overrightarrow{BA} + \overrightarrow{BC} = 2\overrightarrow{BX}$

(b) $\overrightarrow{BA} + \overrightarrow{BC} + \overrightarrow{DA} + \overrightarrow{DC} = 4\overrightarrow{YX}$

(c) $2\overrightarrow{AB} + 2\overrightarrow{BC} + 2\overrightarrow{CA} = \mathbf{0}$

If $\overrightarrow{OA} + \overrightarrow{OB} + \overrightarrow{OC} + \overrightarrow{OD} = 4\overrightarrow{OM}$, find the location of M. (AEB)

10. A hovercraft travelling horizontally in a straight line starts from rest and accelerates uniformly during the first 6 minutes of its journey when it covers 2 km. Then it moves at constant velocity until it experiences a constant retardation which brings it to rest in a further distance of 4 km.

(a) Sketch the velocity–time graph and find the maximum velocity, in km/h attained by the hovercraft.

(b) Determine the time taken during the retardation.

(c) Given that the total journey time is 42 minutes, determine the distance travelled at constant velocity. (AEB)

11. At time $t = 0$ a particle is projected vertically upwards from a point O with speed 19.6 m/s and, two seconds later, a second particle is projected vertically upwards from O with the same speed. Assuming that the only force acting is that due to gravity, express the heights above O of both particles in terms of t and hence, or otherwise, find the value of t when they collide.
Find the speeds of the particles at the instant of collision. (WJEC)

12.

The diagram above shows the speed–time graph for a train which travels from rest in one station to rest at the next station. For each of the time intervals OA, AB and BC, state the value of the train's acceleration.
Calculate the distance between the stations. (AEB)

13. An airport has a straight level runway of length 3000 metres. During take-off, a jet aircraft, starting from rest, moves with constant acceleration along the runway and reaches its take-off speed of 270 km/h after 40 seconds. Find

(a) the acceleration of the jet during take-off in m/s^2

(b) the fraction of the length of the runway used by the jet during its take-off.
(NEAB)

14. A lift travels vertically upwards from rest at floor A to rest at floor B, which is 20 m above A, in three stages as follows. Firstly, the lift accelerates from rest at A at $2 \ m/s^2$ for 2 s; secondly, it travels at a constant speed; thirdly, it slows down uniformly at $4 \ m/s^2$, coming to rest at B.

Sketch the velocity-time graph for this motion, and show that the journey from A to B takes $6\frac{1}{2}$ seconds. (UCLES)

15. A car is moving along a straight road with uniform acceleration. The car passes a check-point A with a speed of 12 m/s and another check-point C with a speed of 32 m/s. The distance between A and C is 1100 m.

(a) Find the time, in s, taken by the car to move from A to C.

Given that B is the midpoint of AC,

(b) find, in m/s to 1 decimal place, the speed with which the car passes B.
(ULEAC)

16. A train moves from rest at a station, Amesbury, and covers the first 1.8 km of its journey with uniform acceleration. It then travels for 18 km at a uniform speed, and then decelerates uniformly for the final 1.2 km to come to rest at Birchfield. The final journey time is 20 minutes.

(a) Sketch the speed–time graph for this journey.

(b) Calculate, in km/h, the maximum speed attained.

(c) Calculate, in km/h^2, the final deceleration. (ULEAC)

17. A motorist, travelling at 120 km/h on a motorway, passes a police speed check point. The motorist immediately decelerates at a rate of $360 \ km/h^2$.

A police car at the speed check point starts from rest at the instant the motorist passes it, accelerates uniformly to a speed of 130 km/h and then travels at this speed. Given that it overtakes the motorist after 3 minutes and then decelerates,

(a) determine the distance from the speed check to the point where the police car overtakes the motorist

(b) find the time (in minutes) during which the police car is travelling at constant speed. (AEB)

18. Three forces $(\mathbf{i}+\mathbf{j})$ N, $(-5\mathbf{i}+3\mathbf{j})$ N and $k\mathbf{i}$ N, where $\mathbf{i}$ and $\mathbf{j}$ are perpendicular unit vectors, act at a point. Express the resultant of these forces in the form $a\mathbf{i}+b\mathbf{j}$ and find its magnitude in terms of k.
Given that the resultant has magnitude 5 N find the two possible values of k.
Take the larger value of k and find the tangent of the angle between the resultant and the unit vector $\mathbf{i}$. (AEB)

19. A vehicle travelling on a straight horizontal track joining two points A and B accelerates at a constant rate of 0.25 m/s² and decelerates at a constant rate of 1 m/s². It covers a distance of 2.0 km from A to B by accelerating from rest to a speed of v m/s and travelling at that speed until it starts to decelerate to rest. Express in terms of v the times taken for acceleration and deceleration.

Given that the total time for the journey is 2.5 minutes find a quadratic equation for v and determine v, explaining clearly the reason for your choice of the value of v. (AEB)

20. O is the origin; A and B are points such that $\overrightarrow{OA} = \mathbf{a}$ and $\overrightarrow{OB} = \mathbf{b}$. M is the midpoint of AB. P is a point on OB such that $OP = 2PB$ and Q is a point on OM such that $OQ = \frac{1}{2}QM$. The line PQ is produced to meet OA at R. Express $\overrightarrow{OR}$ in terms of $\mathbf{a}$ and $\mathbf{b}$. (WJEC)

The instruction for answering questions 21 to 24 is:
if the following statement must *always* be true, write T, otherwise write F.

21. If $\mathbf{F}_1 = 2\mathbf{i}+3\mathbf{j}$ and $\mathbf{F}_2 = 2\mathbf{i}-3\mathbf{j}$ then $\mathbf{F}_1$ and $\mathbf{F}_2$ are equal and opposite.

22. A particle of weight W is on a plane inclined at α to the horizontal. The component of the weight parallel to the plane is $W\cos\alpha$.

23. The resultant of $\overrightarrow{AB}$ and $\overrightarrow{BC}$ is $\overrightarrow{CA}$.

24. Velocity is the rate of increase of distance.

25. In $\triangle OPQ$, $\overrightarrow{OP} = 2\mathbf{p}$, $\overrightarrow{OQ} = 2\mathbf{q}$, and M and N are the midpoints of OP and OQ respectively.

   (a) By expressing the vectors $\overrightarrow{PQ}$ and $\overrightarrow{MN}$ in terms of $\mathbf{p}$ and $\mathbf{q}$, prove that
   $$\overrightarrow{MN} = \tfrac{1}{2}\overrightarrow{PQ}.$$

   The lines PN and QM intersect at the point G.

   (b) Express, in terms of $\mathbf{p}$ and $\mathbf{q}$, the vectors $\overrightarrow{PN}$ and $\overrightarrow{QM}$.

   (c) Given that $\overrightarrow{GN} = \lambda\overrightarrow{PN}$ prove that
   $$\overrightarrow{OG} = 2\lambda\mathbf{p}+(1-\lambda)\mathbf{q}.$$

   (d) Given that $\overrightarrow{GM} = \mu\overrightarrow{QM}$ find $\overrightarrow{OG}$ in terms of μ, $\mathbf{p}$ and $\mathbf{q}$.

   (e) Hence prove that
   $$\overrightarrow{OG} = \tfrac{2}{3}(\mathbf{p}+\mathbf{q}).$$
   (ULEAC)

26. A particle moving in a straight line with speed u m/s is retarded uniformly for 16 seconds so that its speed is reduced to $\frac{1}{4}u$ m/s. It travels at this reduced constant speed for a further 16 seconds. The particle is then brought to rest by applying a constant retardation for a further 8 seconds. Draw a speed–time graph and hence, or otherwise,

(a) express both retardations in terms of u

(b) show that the total distance travelled over the two periods of retardation is $11u$ m

(c) find u given that the total distance travelled in the 40 seconds in which the speed is reduced from u m/s to zero is 45 m. (AEB)

27. An underground train travels along a straight horizontal track from station A to station B. The train accelerates uniformly from rest at A to a maximum speed of 20 m/s, then travels at this speed for 30 seconds before slowing down uniformly to come to rest at B. The acceleration is f m/s², the retardation is $2f$ m/s² and the time for the whole journey is 1 minute. Sketch the velocity–time graph for the journey. Calculate

(a) the distance between the stations A and B

(b) the value of f. (NEAB)

28. A car starts from rest at time $t = 0$ seconds and moves with a uniform acceleration of magnitude 2.3 m/s² along a straight horizontal road. After T seconds, when its speed is V m/s, it immediately stops accelerating and maintains this steady speed until it hits a brick wall when it comes instantly to rest. The car has then travelled a distance of 776.25 m in 30 s.

(a) Sketch a speed–time graph to illustrate this information.

(b) Write down an expression for V in terms of T.

(c) Show that

$$T^2 - 60T + 675 = 0$$ (ULEAC)

29. A railway train is moving along a straight level track with a speed of 10 m/s when the driver sights a signal which is at green. As soon as the signal is sighted the train starts to accelerate. Given that the acceleration has a constant value of f m/s², show that the distance in metres moved by the train during the nth second after the signal is sighted is

$$\left(10 - \frac{f}{2} + nf\right)$$

Find the value of f given that the train travels 25 m during the 8th second after the signal is sighted. (NEAB)

CHAPTER 5

NEWTON'S LAWS OF MOTION

FORCE AND MOTION

Early mathematicians, right up to the Middle Ages, were convinced that whenever a body is moving there must be a force acting on it, i.e. a force is needed to 'make it keep moving'.

We can see, now, that something is wrong with that hypothesis by considering, for example, a puck skimming across the ice rink during an ice hockey match.

The puck is struck with a stick and sent moving (i.e. a force *starts* the motion), but what happens next? The puck continues to move although there is nothing to push it once it has left the hockey stick.

This is just one example where motion exists without a force to cause it.

However, it was not until 1687 that, with the publication of Newton's Laws of Motion, the old hypothesis was discarded and a completely new school of thought established.

NEWTON'S FIRST LAW

This law is the result of one of those brilliant pieces of deduction which from time to time produce an idea so simple that it is difficult to understand why it had eluded thinkers for so long. The law states that:

A body will continue in its state of rest, or of uniform motion in a straight line, unless an external force is applied to it.

This immediately explains the motion of the puck; once struck and set in motion, it continues to move in a straight line until some other force intervenes.

Further deductions can be made from Newton's first law

- If a body is at rest, or is moving with constant velocity, then there is no resultant force acting on it and any forces that do act must balance exactly, i.e. must be in equilibrium.

- If the speed of a moving object is changing, there must be a resultant force acting on it.

- If the *direction* of motion of a moving object is changing, i.e. it is not moving in a straight line, there must be a resultant force acting on it. (So there is always a force acting on a body that is moving in a curve, even if the speed is constant.)

Newton's first law in effect defines what force is, i.e. force is the quantity that, when acting on a body, changes the velocity of that body.

Now if the velocity of a body changes, there is an acceleration, so we can say:

If a body has an acceleration there is a resultant force acting on it.
If a body has no acceleration there is no resultant force acting on it.

A body has no acceleration when it is at rest, *or when it is moving with constant velocity* so it should not be thought that a force is needed to keep a body moving with constant velocity.

Examples 5a

1. A body is at rest under the action of the forces shown, all forces being measured in newtons. Find the values of F and R.

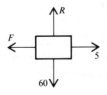

The body is at rest therefore there is no resultant either horizontally or vertically.

Horizontally $5 - F = 0$ $\Rightarrow$ $F = 5$

Vertically $R - 60 = 0$ $\Rightarrow$ $R = 60$

2. A particle of weight 7 N, hanging at the end of a vertical light string is moving upward with constant velocity. Find the tension in the string.

The velocity of the particle is constant so it has no acceleration and therefore the resultant force is zero.

Vertically $\quad T - 7 = 0 \quad\quad \Rightarrow \quad\quad T = 7$

$\therefore \quad$ the tension in the string is 7 N.

3. The diagram shows the forces that act on a particle. Determine whether the resultant acceleration of the particle is horizontal, vertical or in some other direction, if (a) $P = 5$ (b) $P = 8$.

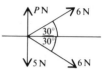

(a) Horizontally the forces do not balance as there is no force to the left, so there is a horizontal acceleration component.

Vertically the forces balance in pairs, so there is no vertical acceleration.

The resultant acceleration is horizontal.

(b) As in (a) there is a horizontal acceleration component.

The vertical components of the forces of 6 N balance but the 5 N and 8 N do not, so there is a vertical acceleration component.

The resultant acceleration is the combination of the two components.

The resultant acceleration is neither horizontal nor vertical but is in some other direction.

EXERCISE 5a

In each question from 1 to 6, the diagram shows the forces, all in newtons, acting on an object which is at rest. Find P and/or Q.

1.

3.

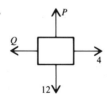

5.

2.

4.

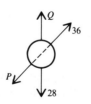

6.

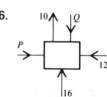

In each question from 7 to 12 determine whether or not the body shown in the diagram has an acceleration. If there is an acceleration state whether it is (i) horizontal (ii) vertical (iii) neither horizontal nor vertical.

7.

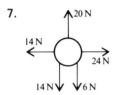

9.

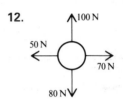

11.

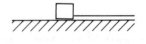

8.

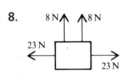

10.

12.

13. The diagram shows a block in rough contact with a horizontal surface. It is being pulled along by a horizontal string.

(a) Make a copy of the diagram and on it mark all the forces acting on the block.

(b) What can you say about the tension in the string compared with the frictional force if the block

(i) is accelerating (ii) moves with constant velocity?

14. The forces, in newtons, acting on a body are $-2\mathbf{i} + 6\mathbf{j}$, $5\mathbf{i} - 3\mathbf{j}$, $4\mathbf{j}$ and $-3\mathbf{i} - 3\mathbf{j}$. Determine whether the body is accelerating and, if it is, state in which direction.

15. The diagram shows a plank with one end in rough contact with the ground and resting in smooth contact with a post. The forces acting on the plank are marked on the diagram.

(a) Collect the vertical components of the forces.

(b) Collect the horizontal components of the forces.

(c) Given that the plank is at rest, form equations by giving a numerical value to the expression you found (i) in part (a) (ii) in part (b).

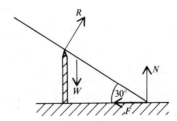

16.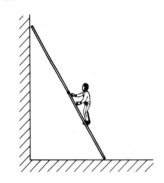

The diagram shows a ladder, with its foot on rough ground, leaning against a smooth wall. The weight of the ladder, W, acts through the midpoint; a man, also of weight W, is standing on the ladder as shown.

(a) Mark all the forces that act on the ladder, on a copy of the diagram.

(b) Write down two expressions involving the forces, that you could use if you were asked to find out whether the ladder is stationary.

17. A wind of strength P newtons is blowing a small boat on a pond. The boat, whose weight is W newtons, is moving with constant velocity. Draw a diagram and mark all the forces acting on the boat, using F for the frictional resistance of the water and R for the supporting force exerted by the water (both in newtons). What is the relationship between

(a) P and F (b) R and W?

NEWTON'S SECOND LAW

This law defines the relationship between force, mass and acceleration. It seems reasonable to accept that

(i) for a body of a particular mass, the bigger the force is, the bigger the acceleration will be

(ii) the larger the mass is, the larger will be the force needed to produce a particular acceleration.

Experimental evidence verifies that the force F is proportional both to the acceleration a and to the mass m.

i.e. $F \propto ma$

or $F = kma$ where k is a constant

Now if $m = 1$ and $a = 1$ then $F = k$, so the amount of force needed to give a mass of 1 kg an acceleration of 1 m/s^2 is given by k.

If this amount of force is chosen as the unit of force we have $k = 1$ and

$$F = ma$$

The unit of force is called the newton (N) and is defined as the amount of force that gives 1 kg an acceleration of 1 m/s^2.

Now we know that acceleration and force are both vector quantities and that mass is scalar. We also know that if $\mathbf{p} = k\mathbf{q}$, then $\mathbf{p}$ and $\mathbf{q}$ are parallel vectors.

Therefore, from the equation $\mathbf{F} = m\mathbf{a}$ we see that:

the vectors $\mathbf{F}$ and $\mathbf{a}$ are parallel, i.e. the direction of an acceleration is the same as the direction of the force that produces it.

When more than one force acts on a body, $\mathbf{F}$ represents the resultant force.

If the force is constant the acceleration also is constant and, conversely, if the force varies, so does the acceleration.

If the acceleration is zero, the resultant force is zero – in other words Newton's first law follows from the second.

To sum up:

> **The resultant force acting on a body of constant mass is equal to the mass of the body multiplied by its acceleration.**
>
> $$\mathbf{F(N)} \;=\; m(\text{kg}) \times \mathbf{a}(\text{m/s}^2)$$
>
> **The resultant force and the acceleration are in the same direction**

Examples 5b

1. A force of 12 N acts on a body of mass 5 kg. Find the acceleration of the body.

Using $F = ma$ gives $12 = 5a$

$\Rightarrow \qquad a = 2.4$

The acceleration is 2.4 m/s^2 in the direction of the force.

2. A set of forces act on a mass of 3 kg and give it an acceleration of 11.4 m/s^2. Find the magnitude of the resultant of the forces.

If $m = 3$ and $a = 11.4$ using $F = ma$ gives

$$F = 3 \times 11.4 = 34.2$$

The resultant force is of magnitude 34.2 N.

(The direction of the resultant is not asked for but it is in the direction of the acceleration which is not given.)

3. Forces $4\mathbf{i} - 7\mathbf{j}$ and $-\mathbf{i} + 3\mathbf{j}$ act on a particle of mass 2 kg. Given that the forces are measured in newtons, express the acceleration of the particle in the form $a\mathbf{i} + b\mathbf{j}$ and find its magnitude. Find the angle between the direction of the acceleration and the vector $\mathbf{i}$.

The resultant force is $\quad 4\mathbf{i} - 7\mathbf{j} + (-\mathbf{i} + 3\mathbf{j}) = 3\mathbf{i} - 4\mathbf{j}$

Using $\mathbf{F} = m\mathbf{a}$ gives $3\mathbf{i} - 4\mathbf{j} = 2\mathbf{a}$

So, measured in m/s², $\qquad \mathbf{a} = \frac{1}{2}(3\mathbf{i} - 4\mathbf{j}) = 1.5\mathbf{i} - 2\mathbf{j}$

The magnitude of $\mathbf{a}$ is

$$|1.5\mathbf{i} - 2\mathbf{j}|$$
$$= \sqrt{(1.5^2 + [-2]^2)} = 2.5$$

The magnitude of the acceleration is 2.5 m/s².

$\tan\alpha = 2/1.5 \quad \Rightarrow \quad \alpha = 53°$ to the nearest degree.

Therefore the direction of the acceleration is at 53° to $\mathbf{i}$ as shown.

4. The diagram shows the forces that act on a particle of mass 5 kg, causing it to move vertically downward with an acceleration a m/s². Find the values of P and a.

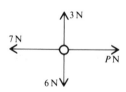

There is no horizontal motion

$\therefore \quad \rightarrow \quad$ gives $P - 7 = 0 \quad \Rightarrow \quad P = 7$

The resultant force vertically downwards is $(6 - 3)$ N $= 3$ N

Using $F = ma \quad \downarrow \quad$ gives

$$3 = 5a \quad \Rightarrow \quad a = \frac{3}{5}$$

In some problems we are given some facts about how an object is *being made to move* with constant acceleration and also other information about the *motion* of the object. So we use both $F = ma$ and one of the equations of motion derived in Chapter 2 which contains a.

5. A force F newtons acts on a particle of mass 3 kg.

    (a) If the particle accelerates uniformly from 2 m/s to 8 m/s in 2 seconds, find the value of F.

    (b) If $F = 6$ find the displacement of the particle 4 seconds after starting from rest.

    (a) For the motion of the particle we have:

$$u = 2, \quad v = 8, \quad t = 2 \text{ and } a \text{ is required.}$$

Using $v = u + at$ gives $8 = 2 + 2a$

$\Rightarrow \qquad a = 3$

Now using $F = ma$ gives $F = 3 \times 3$

$\Rightarrow \qquad F = 9$

The force is 9 N.

    (b) This time we know the force so we use Newton's Law first.

Using $F = ma$ gives $6 = 3a$ $\Rightarrow$ $a = 2$

Now for the motion of the particle:

$$u = 0, \quad a = 2, \quad t = 4 \text{ and } s \text{ is required.}$$

Using $s = ut + \frac{1}{2}at^2$ gives $s = 0 + \frac{1}{2}(2)(4)^2$

$\Rightarrow \qquad s = 16$

The displacement of the particle is 16 m.

EXERCISE 5b

1. A force of 12 N acts on a body of mass 8 kg. What is the acceleration of the body?

2. The acceleration of a particle of mass 2 kg is 14 m/s². What is the resultant force acting on the particle?

3. A force of 420 N acts on a block, causing an acceleration of 10.5 m/s². Assuming that no other force acts on the block, find its mass.

4. A force, measured in newtons, is represented by $5\mathbf{i} - 12\mathbf{j}$. If the force acts on an object of mass 26 kg find, in the form $a\mathbf{i} + b\mathbf{j}$, the acceleration of the object. What is the magnitude of the acceleration?

5. Find, in Cartesian vector form, the force which acts on an object of mass 5 kg, producing an acceleration of $7\mathbf{i} + 2\mathbf{j}$ measured in m/s².

6. In each diagram the forces shown (measured in newtons) cause a body of mass 8 kg to move with the acceleration shown. Find P and/or Q in each case.

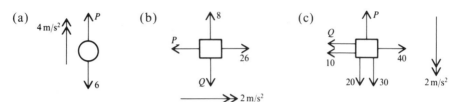

(a)   (b)   (c)

7. Each diagram shows the forces acting on a body of mass 3 kg. Find the magnitude and direction of the acceleration of the particle in each case.

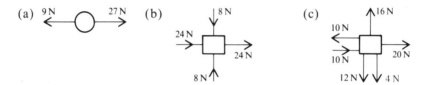

(a)   (b)   (c)

8. In each diagram the mass of the body is m kilograms. Find m and P.

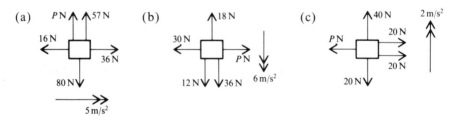

(a)   (b)   (c)

9. A particle of mass 3 kg is accelerating vertically downwards at 5 m/s² under the action of the forces shown, all measured in newtons. Find the values of P and Q.

10. A body of mass 2 kg accelerates uniformly from rest to 16 m/s in 4 seconds. Find the resultant force acting on the body.

11. A force of 100 N acts on a particle of mass 8 kg. If the particle is initially at rest, find how far it travels in the first 5 seconds of its motion.

12. A block of mass 6 kg is pulled along a smooth horizontal surface by a horizontal string. If the block reaches a speed of 20 m/s in 4 seconds from rest, find the tension in the string.

13. Forces of $2\mathbf{i} + 3\mathbf{j}$ and $-6\mathbf{i} + 7\mathbf{j}$ act on a body of mass 10 kg. Given that the forces are measured in newtons find in the form $a\mathbf{i} + b\mathbf{j}$

   (a) the acceleration of the body

   (b) the velocity of the body 2 seconds after starting from rest.

14. A constant force of 80 N acts for 7 seconds on a body, initially at rest, giving it a velocity of 35 m/s. Find the mass of the body.

15. A force of 12 N acts on a particle of mass 60 kg causing the velocity of the particle to increase from 3 m/s to 7 m/s. Find the distance that the particle travels during this period.

16. A body of mass 120 kg is moving in a straight line at 8 m/s when a force of 40 N acts in the direction of motion for 18 seconds. What is the speed of the body at the end of this time?

WEIGHT

Consider an object of mass m kg, falling freely under gravity with an acceleration g m/s^2.

We know that the force producing the acceleration is the weight, W newtons, of the object, so using $F = ma$ gives

$$W = mg$$

i.e.　　**a body of mass m kilograms has a weight of mg newtons.**

For example, taking the value of g as 9.8,

the weight, W newtons, of a person whose mass is 55 kg is given by

$$W = mg = 55 \times 9.8 = 539$$

therefore the person's weight is 539 N.

If the weight of a rockery stone is 1078 N, its mass, m kg, is given by

$$W = mg \quad \Rightarrow \quad m = \frac{W}{g} = \frac{1078}{9.8} = 110$$

therefore the mass of the stone is 110 kg.

You can get an idea of what a newton is like if you think of the weight of a smallish apple.

Example 5c

A rope with a bucket attached to the end is used to raise water from a well. The mass of the empty bucket is 1.2 kg and it can raise 10 kg of water when full. Taking g as 9.8 find the tension in the rope when

(a) the empty bucket is lowered with an acceleration of 2 m/s²

(b) the full bucket is raised with an acceleration of 0.3 m/s²

(a) The acceleration of the bucket is downward so the resultant force acts downwards.
The weight of the bucket is $1.2g$ N $= 1.2 \times 9.8$ N
The resultant force downward is $(1.2g - T)$ N

Using $F = ma \downarrow$ gives

$$1.2 \times 9.8 - T = 1.2 \times 2$$
$$T = 1.2 \times 9.8 - 1.2 \times 2$$
$$= 9.36$$

The tension in the rope is 9.36 N.

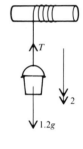

(b) The acceleration of the full bucket is upward so the resultant force acts upwards.
The weight of the full bucket is 11.2×9.8 N
Resultant force $\uparrow$ is $(T - 11.2g)$ N

Using $F = ma \uparrow$ gives

$$T - 11.2 \times 9.8 = 11.2 \times 0.3$$
$$T = 11.2 \times 0.3 + 11.2 \times 9.8$$
$$= 113.12$$

The tension in the rope is 113 N (3 sf).

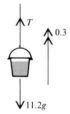

Note that readers who are competent in algebraic factorising may prefer to give the line where T is calculated in part (a) as $T = 1.2 (9.8 - 2) = 9.36$ and the similar line in part (b) as $T = 11.2 (9.8 + 0.3) = 113.12$.

EXERCISE 5c

In this exercise take the value of g on earth as 9.8.

1. (a) Find the weight of a body of mass 5 kg.

(b) What is the mass of a sack of potatoes of weight 147 N?

(c) What is the weight of a tennis ball of mass 60 g?

2. On the moon the acceleration due to gravity is 1.2 m/s². What answers to question 1 would a student in a lunar school give?

3. A particle of mass 2 kg, attached to the end of a vertical light string, is being pulled up, with an acceleration of 5.8 m/s², by the string.

   Find the tension in the string.

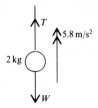

4. A mass of 6 kg is moving vertically at the end of a light string. Find the tension in the string when the mass has an acceleration of

   (a) 5 m/s² downwards   (b) 7 m/s² upwards   (c) zero.

5. The tension in a string, which has a particle of mass m kilograms attached to its lower end, is 70 N.

   Find the value of m if the particle has

   (a) an acceleration of 3 m/s² upwards

   (b) an acceleration of 9 m/s² downwards

   (c) a constant velocity of 4 m/s upwards

   (d) a constant velocity of 4 m/s downwards.

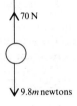

6. A goods lift with a mass of 750 kg can be raised and lowered by a cable. The maximum load it can hold is 1000 kg.

   (a) Find the tension in the cable when

       (i) raising the fully-loaded lift with an acceleration of $\frac{1}{2}$ m/s²

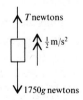

       (ii) lowering the empty lift with an acceleration of $\frac{3}{4}$ m/s²

   (b) The tension in the cable is 14 700 N when the lift, partly loaded, is being raised at constant speed. Find the mass of the load.

7. A balloon of mass 1400 kg is descending vertically with an acceleration of 2 m/s². Find the upward force being exerted on the balloon by the atmosphere.

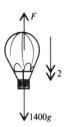

8. A block of mass 4 kg is lying on the floor of a lift that is accelerating at 5 m/s². Find the normal reaction exerted on the block by the lift floor if the lift is

   (a) going up   (b) going down.

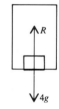

MOTION CAUSED BY SEVERAL FORCES

We have seen that if an object is moving in a particular line, we can analyse the motion by collecting the forces in that direction. However, in the examples so far considered in this chapter the given forces have all been either vertical or horizontal so that collecting them was simple arithmetic. Now we must look at more realistic situations where forces in various directions act on an object.

Examples 5d

1. A body of mass 3 kg is sliding down a smooth plane inclined at 30° to the horizontal.

   (a) Find the acceleration of the mass in terms of g.

   (b) Show that the normal reaction exerted by the plane on the mass is given by $\dfrac{3\sqrt{3}g}{2}$.

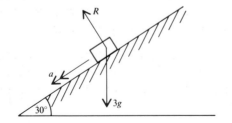

The diagram shows all the forces that act on the body. The body can move in only one way, i.e. down the plane, so the acceleration is marked in that direction. The acceleration is caused only by the components of force that act along the plane so these must be found and collected.

(a) The resultant force $\nearrow$ is $3g \sin 30°$, i.e. $3g(\frac{1}{2})$

    Using $F = ma$ gives $\quad (\frac{3}{2})g = 3a \quad \Rightarrow \quad a = \frac{1}{2}g$

(b) As there is no acceleration perpendicular to the plane, the resultant of the forces in that direction is zero.

    Resolving $\nwarrow$ gives $\quad R - 3g \cos 30° = 0$

$$R = 3g \left(\frac{\sqrt{3}}{2} \right) = \frac{3\sqrt{3}g}{2}$$

Note that in this example the values of $\sin 30°$ and $\cos 30°$ are used in their exact form (using square roots). Readers who have not yet encountered these values should now learn them as they often occur in problems where exact expressions are required. Taking $\cos 30°$ as 0.8860, is correct only to 4 dp and using it can cause difficulty in a problem where exact values are required.

Using Pythagoras' theorem in a sketch of half an equilateral triangle with sides of length 2 units, shows clearly the sine and cosine of both 30° and 60°, as well as the tangent of each angle, and helps in remembering the exact forms.

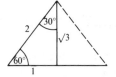

Another angle whose trig. ratios can be given exactly is 45°. For this one a right-angled isosceles triangle shows the values.

The Form in which Answers should be Given

There is no precise value for g so if a numerical value of g is used in a problem all calculations based on that value are approximations.
Further approximations are made if numerical values of trig ratios are introduced (e.g. $\sin 50° = 0.7660$ correct to 4 dp), so answers based on these approximations should be given only to 2 or 3 significant figures.

For these reasons answers are often given in an exact form, i.e. a quantity which has no exact numerical value is left as a symbol such as g, W, m etc. The trig ratios for angles of 30°, 45° and 60° can be expressed in exact surd form and others may be left as, say, $\sin 20°$.

Unless other instructions are given, answers should be presented in exact form. When answers are required to a given degree of accuracy, take the value of g as 9.8 m/s² unless a different value is specified.

2. A cyclist is riding up a hill inclined at 20° to the horizontal. His speed at the foot of the hill is 10 m/s but after 30 seconds it has dropped to 4 m/s. The total mass of the cyclist and his machine is 100 kg and there is a wind of strength 15 N down the slope. Find, corrected to 3 significant figures, the constant driving force exerted by the cyclist up the slope.

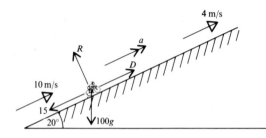

First we will find the acceleration of the cyclist up the slope.

For the motion up the slope:

Known $u = 10$ Required a

             $v = 4$

             $t = 30$

Using $v = u + at$ gives $4 = 10 + 30a$

$\therefore$ $a = -\frac{6}{30} = -\frac{1}{5}$

The cyclist has an acceleration of $-\frac{1}{5}$ m/s^2

      Resolving   ↗   gives $D - 15 - 100g \sin 20°$

Using $F = ma$ ↗   gives $D - 15 - 100g \sin 20° = 100a$

i.e. $D = 15 + 980(0.3420) + 100(-\frac{1}{5})$

                                $= 330.1\ldots$

The cyclist's driving force is 330 N (3 sf).

EXERCISE 5d

1. The diagram shows a small block of mass 2 kg being pulled up a plane inclined at 30° to the horizontal. The block has an acceleration of 0.5 m/s^2.

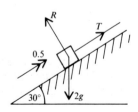

Find an exact expression in terms of g for the tension in the string if

(a) the plane is smooth

(b) the plane is rough and exerts a frictional force of 4 N.

2. A truck is being pulled along a horizontal track by two cables, against
resistances totalling 1100 N, with an acceleration of 0.8 m/s². One cable is
horizontal and the other is inclined at 40° to the track. The tensions in the cables
are shown on the diagram. Taking g as 9.8 find, corrected to 3 significant figures,

(a) the mass of the truck

(b) the vertical force exerted by the track on the truck.

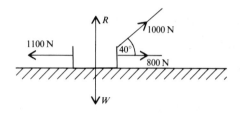

3. Ballast of mass 20 kg is dropped from a balloon that is moving horizontally
with a constant speed of 2 m/s.

(a) Mark on a diagram the forces acting on the ballast as it falls.

(b) What is (i) the vertical component (ii) the horizontal component
of the acceleration of the ballast?

(c) Without doing any further calculation, sketch *roughly* the path of the ballast
as it falls.

4. The diagram shows a small block of mass 5 kg being pulled along a rough
horizontal plane by a string inclined at 60° to the plane. There is a frictional
force of 18 N.

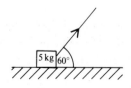

Copy the diagram and on it mark all the forces that act on the block. If the
block has an acceleration of 3 m/s², find the tension in the string and show that
the normal reaction exerted by the plane on the block can be expressed
as $(5g - 33\sqrt{3})$ N.

5.

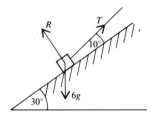

A body of mass 6 kg is sliding down a smooth plane inclined at 30° to the horizontal. Its speed is controlled by a rope inclined at 10° to the plane as shown; the tension in the rope is 10 N. Given that the body starts from rest, find how far down the plane it travels in 5 seconds. Taking *g* as 9.8, give the answer corrected to 2 significant figures.

NEWTON'S THIRD LAW

The statement of this law is:

Action and Reaction are Equal and Opposite

This means that if a body A exerts a force on a body B then B exerts an equal force in the opposite direction on A. This is true whether the two bodies are in contact or are some distance apart, whether they are moving or are stationary.

Consider, for example, a mass resting in a scale pan. The scale pan is exerting an upward force on the mass and the mass is exerting an equal force downward on the scale pan.

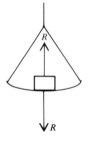

Now consider two particles connected by a taut string. The objects are not in direct contact but exert equal and opposite forces on each other by means of the equal tensions in the string which act inwards at each end.

This is true even if the string passes round a *smooth* body, such as a pulley, which changes the direction of the string.

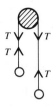

The tensions in the two portions of the string are the same and each portion exerts an inward pull at each end. So in each portion the tension at one end acts on the particle and at the other end the tension acts on the pulley; all these tensions are equal.

(If the string passes round a rough surface the tensions in the different portions of the string are not equal, but the study of this situation is beyond the scope of this book.)

EXERCISE 5e

For each question copy the diagram, making your copy at least twice as big, and mark on it all the forces that are acting on each body. Use either a different colour or a different type of line (e.g. broken and solid) for the forces that act on separate objects. Ignore forces that act on fixed surfaces – these are indicated in the usual way by hatching.

1.

A load hangs from a beam which is supported at each end.

3.

A mass B hangs by a string from another mass A which hangs from a fixed point.

5.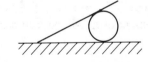

A rod with one end on rough ground rests against an oil drum.

2.

A mass on a table is linked by a string to a mass hanging over the smooth edge.

4.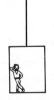

A passenger is in a lift that is being drawn up by a cable.

6.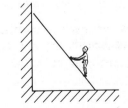

A workman is standing on a rung of a ladder that rests on rough ground and a smooth wall.

THE MOTION OF CONNECTED BODIES

In this work we use a number of conventions that are not completely realistic but are useful in providing simple mathematical models to work with.

For instance we will refer to an inextensible (or inelastic) string, i.e. a string whose length cannot alter, whereas in reality no string is completely inextensible. We shall also be dealing with *smooth* pulleys, i.e. those whose bearings and rim are completely without friction; in practice no pulley is completely smooth.

Having said this however, models based on these conventions can be fairly close to reality.

Consider two particles A and B, connected by a light inextensible string passing over a smooth fixed pulley and suppose that A is heavier than B.

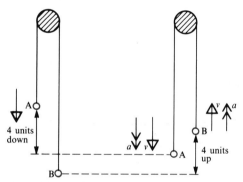

As the particles move, one of them moves upwards in the same way as the other moves downwards, i.e.

the upward speed of B is equal to the downward speed of A,

the upward acceleration of B is equal to the downward acceleration of A.

the distance B moves up is equal to the distance A moves down.

The way in which each particle moves is determined by the forces which act on *that particle alone,* so to analyse the motion we consider the particles separately.

Examples 5f

1. A light inextensible string passes over a smooth fixed pulley and carries particles of masses 5 kg and 7 kg, one at each end. If the system is moving freely, find in terms of g

   (a) the acceleration of each particle

   (b) the tension in the string

   (c) the force exerted on the pulley by the string.

The two particles have the same acceleration, a m/s², and the two parts of the string have the same tension, T newtons.

For each particle we will find the resultant force in the direction of motion and use it in the equation of motion, $F = ma$, in that direction.

$5g = 0.005 kg$

$0.005 g$

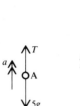
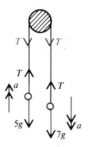

For A: ↑

     The resultant force   is $T - 5g$

     Using $F = ma$ gives $T - 5g = 5a$ [1]

For B: ↓

     The resultant force   is $7g - T$

     Using $F = ma$ gives $7g - T = 7a$ [2]

(a)   Adding [1] and [2] gives $2g = 12a$ $\Rightarrow$ $a = \frac{1}{6}g$

     The acceleration of each particle is $\frac{g}{6}$ m/s².

(b)   From [1], $T - 5g = 5\left(\frac{g}{6}\right)$ $\Rightarrow$ $T = \frac{35g}{6}$

     The tension in the string is $\frac{35g}{6}$ N.

(c)   The string exerts a downward pull on each side of the pulley. Therefore the resultant force exerted on the pulley by the string is $2T$ downwards.

     i.e. $\frac{35g}{3}$ N downwards.

2. A block of mass 6 kg rests on a table top and is connected by a light inextensible string that passes over a smooth pulley, fixed on the edge of the table, to another block of mass 5 kg which is hanging freely. Find, in terms of g, the acceleration of the system and the tension in the string if

     (a)   the table is smooth

     (b)   the table is rough and exerts a frictional force of $2g$ N.

We will use the equation of motion for each block in its direction of motion.

(a)

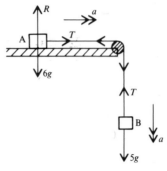

For the block A   → $T = 6a$ [1]

For the block B   ↓ $5g - T = 5a$ [2]

$[1] + [2]$ ⇒ $5g = 11a$

                   ⇒ $a = \dfrac{5g}{11}$

and $T = 6a = \dfrac{30g}{11}$

The acceleration is $\dfrac{5g}{11}$ m/s² and the tension is $\dfrac{30g}{11}$ N.

(b)

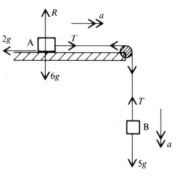

For the block A   → $T - 2g = 6a$ [3]

For the block B   ↓ $5g - T = 5a$ [4]

$[3] + [4]$ ⇒ $3g = 11a$

                   ⇒ $a = \dfrac{3g}{11}$

and $T = 6a + 2g = \dfrac{40g}{11}$

The acceleration is $\dfrac{3g}{11}$ m/s² and the tension is $\dfrac{40g}{11}$ N.

Sometimes the quantities involved in a problem are given only as symbols without units. In such cases no units can be given for the answers.

3. A tractor of mass M is pulling a trailer of mass M_1. If the tractor exerts a steady driving force D find, in terms of M, M_1 and D,

(a) the acceleration of the trailer

(b) the tension in the tow rope.

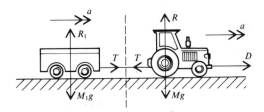

We can treat the tractor and trailer as small blocks with no resistance to motion. The tractor and trailer have the same acceleration.

Considering the trailer, the only force acting → is T

$F = ma \rightarrow$ gives $T = M_1 a$ [1]

Considering the tractor, the force acting → is $D - T$

$F = ma \rightarrow$ gives $D - T = Ma$ [2]

(a) Add [1] and [2] $D = (M + M_1)a$

The acceleration of the trailer is $\dfrac{D}{M + M_1}$

(b) From [1] $T = M_1 \times \dfrac{D}{M + M_1}$

The tension in the tow rope is $\dfrac{M_1 D}{M + M_1}$

4. A particle P of mass $2m$ rests on a rough plane inclined at $30°$ to the horizontal. The frictional force is equal to one-half of the normal reaction. P is attached to one end of a light inelastic string which passes over a smooth pulley fixed at the top of the plane and carries a particle Q of mass $3m$ hanging freely at the other end. Find in terms of g

(a) the normal reaction between P and the plane

(b) the acceleration of P

(c) the force exerted by the string on the pulley.

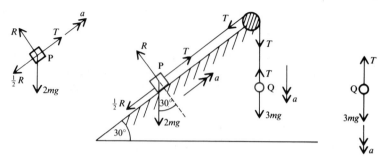

(a)   P does not move perpendicular to the plane so the resultant force in this direction is zero.

For P, resolving ⬉ gives $R - 2mg \cos 30° = 0$

⇒ $R = 2mg \left(\sqrt{3}/2\right) = mg\sqrt{3}$

The normal reaction is $mg\sqrt{3}$

(b)   For P resolving ⬈ gives $T - \frac{1}{2}R - 2mg \sin 30° = 2ma$

i.e. $T - \frac{1}{2}mg\sqrt{3} - 2mg\left(\frac{1}{2}\right) = 2ma$

⇒ $T - \frac{1}{2}mg\sqrt{3} - mg = 2ma$ [1]

For Q resolving ↓ gives $3mg - T = 3ma$ [2]

Add [1] and [2] $mg\left(2 - \frac{1}{2}\sqrt{3}\right) = 5ma$

⇒ $5a = \frac{1}{2}g\left(4 - \sqrt{3}\right)$

The acceleration is $\dfrac{g}{10}\left(4 - \sqrt{3}\right)$

(c)   From [2] $T = 3mg - 3ma$

$$= 3mg - \frac{3mg}{10}\left(4 - \sqrt{3}\right)$$

$$= \frac{3mg}{10}\left(6 + \sqrt{3}\right)$$

The string exerts two equal tensions on the pulley so the
resultant force on the pulley is midway between these
tensions,
i.e. it bisects the angle of 60°.

Resolving in this direction gives $2T \cos 30°$

i.e. $2T\left(\frac{\sqrt{3}}{2}\right)$ ⇒ $T\sqrt{3}$

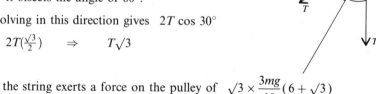

∴   the string exerts a force on the pulley of $\sqrt{3} \times \dfrac{3mg}{10}\left(6 + \sqrt{3}\right)$

i.e. $\dfrac{9mg}{10}\left(2\sqrt{3} + 1\right)$

5. A large box of mass 15 kg can be raised and lowered by a crane. The box contains a load of mass 20 kg. Find
 (i) the tension in the cable of the crane
 (ii) the force exerted by the load on the bottom of the box when the box is accelerating at 2 m/s² (a) upwards   (b) downwards.
  (Take $g = 10$ and give answers corrected to 2 significant figures)

(a)

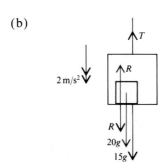   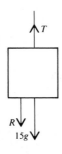   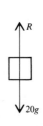

                            Forces acting on box   Forces acting on load

Resolving ↑   for the load $R - 20g = 20 \times 2$
                                     $\Rightarrow$ $R = 240$
              for the box $T - R - 15g = 15 \times 2$
                                     $\Rightarrow$ $T = 420$

 (i) The tension is 420 N.   (ii) The force on the box is 240 N.

(b)

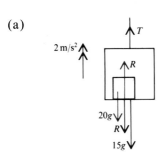   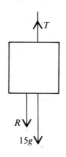   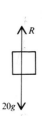

                            Forces acting on box   Forces acting on load

Resolving ↓   for the load $20g - R = 20 \times 2$
                                     $\Rightarrow$ $R = 160$
              for the box $R + 15g - T = 15 \times 2$
                                     $\Rightarrow$ $T = 280$

 (i) The tension is 280 N.   (ii) The force on the box is 160 N.

EXERCISE 5f

Give the answers in terms of *g*.

1. Each diagram shows the forces, all in newtons, acting on two particles connected by a light inextensible string which passes over a fixed smooth pulley. In each case find the acceleration of the system and the tension in the string.

(a)
T↓ T↓

T↑ T↑

○ 2 kg

6 kg ○ ↓ 2g

6g ↓

(b)
T↓ T↓

T↑ T↑

10 kg ○ ○ 15 kg

10g ↓ 15g ↓

(c)
T↓ T↓ a↗

a↓ T↑ T↑

M kg ○ ○ m kg

Mg ↓ mg ↓

2. Two particles are connected by a light inextensible string which passes over a fixed smooth pulley. Find the acceleration of the system and the tension in the string if the masses of the particles are

(a) 5 kg and 10 kg   (b) 12 kg and 8 kg   (c) 2*M* and *M*.

3. A particle of mass 4 kg rests on a smooth plane inclined at 60° to the horizontal. The particle is attached to one end of a light inelastic string which passes over a fixed smooth pulley at the top of the plane and carries a particle of mass 2 kg at the other end. Find

(a) the acceleration of the system   (b) the tension in the string.

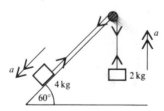

4. Two particles of masses 8 kg and 4 kg hang one at each end of a light inextensible string which passes over a fixed smooth pulley. Find

(a) the acceleration of the system when the particles are released from rest

(b) the distance that each particle moves during the first 5 seconds.

5. A particle of mass 5 kg rests on a smooth horizontal table and is attached to one end of a light inelastic string. The string passes over a fixed smooth pulley at the edge of the table and a particle of mass 3 kg hangs freely at the other end. When the system is released from rest find

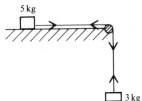

(a) the acceleration of the system

(b) the tension in the string.

6. A car of mass 800 kg exerting a driving force of 2.2 kN (i.e. 2200 N) is pulling a trailer tent of mass 300 kg along a level road. If there is no resistance to the motion of either the car or the trailer find the acceleration of the car and the tension in the towbar.

7. A lift of mass 800 kg is operated by a cable as shown in the diagram. A passenger of mass 70 kg is standing in the lift. Find, stating what object you can use to represent the passenger,

(i) the force exerted by the passenger on the floor of the lift
(ii) the tension in the cable
when the lift is accelerating

(a) upwards   (b) downwards at 0.5 m/s.

(Draw separate diagrams for the lift and for the passenger.)

8. Two particles A and B rest on the smooth inclined faces of a fixed wedge. The particles are connected by a light inextensible string that passes over a fixed smooth pulley at the vertex of the wedge as shown in the diagram. If A and B are each of mass 4 kg find the force exerted by the string on the pulley when the system is moving freely.

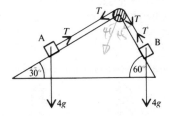

9. Two particles of masses 2 kg and 6 kg are attached one to each end of a long light inextensible string which passes over a fixed smooth pulley. The system is released from rest and the heavier particle hits the ground after 2 seconds. Find the height of this particle above the ground when it was released, and the speed at which it hits the ground.

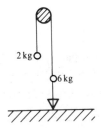

10.

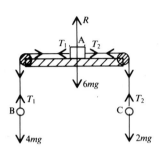

The diagram shows a block A of mass $6m$ resting on a smooth horizontal table. A light inelastic string passes over a fixed smooth pulley at one edge of the table and connects A to a block B of mass $4m$. Another similar string passes over a smooth pulley fixed at the opposite edge of the table and carries a block C of mass $2m$. Assuming that all the moving parts are in a vertical plane find

(a)   the acceleration of the system

(b)   the tension in each string.

(Draw separate diagrams for A, B and C.)

11. A lift of mass 500 kg carrying a load of 80 kg is drawn up by a cable. The lift first accelerates at $g/12$ m/s^2 from rest to its maximum speed which is maintained for a time, after which the lift decelerates to rest at $g/10$ m/s^2. For each of these three stages of motion find

(a)   the tension in the cable

(b)   the force exerted by the load on the floor of the lift.

PROBLEMS IN WHICH THE MOTION CHANGES

An aspect of work on pulley systems that has not yet been considered is what happens when a string breaks or goes slack. This situation is covered in the next example.

Example 5g

Two particles of masses 1 kg and 3 kg are attached to the ends of a long light inelastic string which passes over a fixed smooth pulley. The system is held with both particles hanging at a height of 2 m above the ground, and is released from rest. In the ensuing motion the heavier particle hits the ground and does not rebound. Find the greatest height reached by the mass of 1 kg.

Until the 3 kg mass hits the ground, the masses move as a simple connected system.

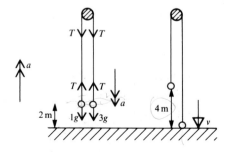

For the 1 kg mass $\uparrow$ $T - 1g = 1a$

For the 3 kg mass $\downarrow$ $3g - T = 3a$

Adding gives $2g = 4a$ $\Rightarrow$ $a = \frac{1}{2}g$

For the motion of the 3 kg mass over a distance of 2 m we have $u = 0$, $a = \frac{1}{2}g$, $s = 2$ and we want the final speed v, which is the speed of each mass at the moment of impact.

Using $v^2 - u^2 = 2as$ gives
$$v^2 - 0 = 2 \times \tfrac{1}{2}g \times 2 \qquad \Rightarrow \qquad v = \sqrt{2g}$$

Once the 3 kg mass hits the ground the string becomes slack and no longer exerts any tension on the 1 kg mass. This mass therefore moves on upwards, with an initial speed of $\sqrt{2g}$, under the action of its weight alone, i.e. with an acceleration of $-g$ upwards.

When it reaches its greatest height,
its speed is zero.
For this part of the motion:

$u = \sqrt{2g}$, $a = -g$, $v = 0$

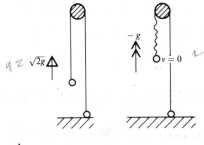

Using $v^2 - u^2 = 2as$ gives
$$0 - 2g = 2(-g)s \qquad \Rightarrow \qquad s = 1$$

The 1 kg mass rises a distance of 1 m above its position when the string went slack. As this point was already 4 m above ground level, the greatest height reached by the 1 kg mass is 5 m above the ground.

EXERCISE 5g

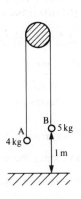

1. Two particles, A of mass 4 kg and B of mass 5 kg, are connected by a light inextensible string passing over a smooth pulley. Initially B is 1 m above a fixed horizontal plane. If the system is released from rest in this position, find

   (a) the acceleration of each particle

   (b) the speed of each particle when B hits the plane

   (c) the further time during which A continues to rise
        (assuming that it does not reach the pulley).

2. Two particles P and Q are connected by a long, light inextensible string passing over a smooth pulley. The mass of P is m, the mass of Q is $2m$ and the particles are held so that each is 3 m below the pulley. The system is released from rest and after 1 second the string breaks. Find

(a) the speed of each particle at that instant

(b) the further distance that P rises.

3.

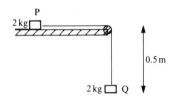

The diagram shows a particle P lying in contact with a smooth table top 1.5 m above the floor. A light inextensible string of length 1 m connects P to another particle Q hanging freely over a small smooth pulley at the edge of the table. The mass of each particle is 2 kg, and P is held at a point distant 0.5 m from the edge of the table. When the system is released from rest find

(a) the speed of each particle when P reaches the edge of the table

(b) the tension in the string.

4. If, in question 3, P slips over the pulley without any change in its speed find, for the subsequent motion,

(a) the acceleration of the system

(b) the tension in the string.

In questions 5 to 8 use $g = 10$ and give answers corrected to 2 significant figures.

5. A particle A of mass $10m$ lies on a horizontal table and is connected by a light inextensible string to a particle B of mass $8m$. The string passes over a smooth pulley fixed at the edge of the table and B hangs freely. The table is rough and exerts a frictional force of magnitude $2mg$ on A.

(a) Find the acceleration of the system.

(b) If the system is released from rest when A is 1.2 m from the pulley find, in terms of g, the speed of the particles when A reaches the pulley.

6. If in question 5 the string snaps when A is 0.6 m from the edge of the table

(a) state the way in which B now moves

(b) find the speed with which A reaches the pulley.

7. A car of mass 1 tonne is pulling a caravan of mass 800 kg along a level straight road. There is a total resistance to the motion of 450 N; the individual resistances on the car and caravan are in the ratio of their masses. If the combination accelerates uniformly from rest to 20 m/s in $12\frac{1}{2}$ seconds find

   (a) the tension in the tow bar

   (b) the driving force exerted by the car's engine.

8. If in question 7 the tow bar snaps at the instant when the speed reaches 20 m/s and the car continues with the same driving force, find

   (a) the subsequent acceleration of the car

   (b) the deceleration of the caravan

   (c) how long it takes for the caravan to stop.

CHAPTER 6

FORCES IN EQUILIBRIUM. FRICTION

CONCURRENT FORCES IN EQUILIBRIUM

A body that is at rest, or is moving with constant velocity, is in a state of equilibrium.

The acceleration of a body in equilibrium is zero in any direction therefore the resultant force in any direction is also zero.

The converse of this statement is not necessarily true because, although forces with zero resultant cannot make an object move in a line they can, as we shall see later on, cause an object to *turn,* e.g.

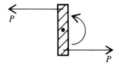

However a set of *concurrent* forces (i.e. all passing through one point) can never cause turning so, as at present we will deal only with concurrent forces, the problem of turning will not arise yet.

We saw in Chapter 3 that the resultant of a set of forces can be found by collecting components in each of two perpendicular directions, giving

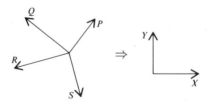

Now if the resultant is zero, the collected components in each direction must individually be zero, i.e. $X = 0$ and $Y = 0$.

Applying this fact to a concurrent system in equilibrium, in which some forces are unknown, provides a method for finding the unknown quantities.

Examples 6a

1. A particle of weight 16 N is attached to one end of a light string whose other end is fixed. The particle is pulled aside by a horizontal force until the string is at 30° to the vertical. Find the magnitudes of the horizontal force and the tension in the string.

Let P newtons and T newtons be the magnitudes of the horizontal force and the tension respectively.

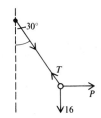

Resolving	→	gives	$P - T \sin 30° = 0$	
i.e.			$P - T \times \frac{1}{2} = 0$	[1]
Resolving	↑	gives	$T \cos 30° - 16 = 0$	
i.e.			$T \times \frac{1}{2}\sqrt{3} - 16 = 0$	[2]

From [2] $T = \frac{32}{\sqrt{3}} = \frac{32\sqrt{3}}{3}$

From [1] $P = \frac{1}{2}T = \frac{16\sqrt{3}}{3}$

Therefore the magnitude of the horizontal force is $\frac{16\sqrt{3}}{3}$ N and the magnitude of the tension is $\frac{32\sqrt{3}}{3}$ N.

2. A load of mass 26 kg is supported in equilibrium by two ropes inclined at 30° and 60° to the horizontal as shown in the diagram. Find in terms of g the tension in each rope.

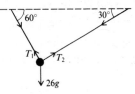

The tensions in the ropes act in perpendicular directions so we will resolve in these directions.

Let the tensions in the ropes be T_1 and T_2 newtons.

Resolving   ↖ $T_1 - 26g \cos 30° = 0$

∴ $T_1 = 26g(\frac{\sqrt{3}}{2}) = 13g\sqrt{3}$

Resolving   ↗ $T_2 - 26g \sin 30° = 0$

∴ $T_2 = 26g(\frac{1}{2}) = 13g$

The tensions in the ropes are $13g\sqrt{3}$ N   and $13g$ N.

3. In each diagram the forces are measured in newtons and are in equilibrium. Find the values of P and Q.

(a)   (b)

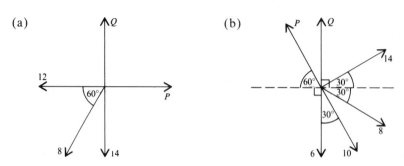

(a) Resolving horizontally and vertically gives simple equations.

Resolving   → $P - 12 - 8 \cos 60° = 0$

⇒ $P - 12 - 8 \times \frac{1}{2} = 0$

∴ $P = 16$

Resolving   ↑ $Q - 14 - 8 \sin 60° = 0$

⇒ $Q - 14 - 8(\frac{\sqrt{3}}{2}) = 0$

∴ $Q = 14 + 4\sqrt{3}$

(b) If we resolve perpendicular to P, the equation we get does not contain P and so gives the value of Q. Similarly, to find P easily we can resolve perpendicular to Q.

Resolving ⊥ to P ↗

$14 + Q \cos 60° + 8 \cos 60° - 6 \cos 60° = 0$

⇒ $14 + \frac{1}{2}Q + \frac{1}{2}(8 - 6) = 0$

⇒ $Q = -30$

Resolving ⊥ to Q →

$14 \cos 30° + 8 \cos 30° + 10 \cos 60° - P \cos 60° = 0$

$(14 + 8)(\frac{\sqrt{3}}{2}) + 10(\frac{1}{2}) - \frac{1}{2}P = 0$

⇒ $P = 22\sqrt{3} + 10$

If the forces are given in the form $a\mathbf{i} + b\mathbf{j}$ then, because a and b are the magnitudes of components in the directions of $\mathbf{i}$ and $\mathbf{j}$, the sum of the coefficients of $\mathbf{i}$ is zero and similarly for $\mathbf{j}$.

4. Forces $2\mathbf{i} - 3\mathbf{j}$, $7\mathbf{i} + 4\mathbf{j}$, $-5\mathbf{i} - 9\mathbf{j}$, $P\mathbf{i} + 2\mathbf{j}$, and $\mathbf{i} - Q\mathbf{j}$, are in equilibrium. Find the values of P and Q.

$$(2\mathbf{i} - 3\mathbf{j}) + (7\mathbf{i} + 4\mathbf{j}) + (-5\mathbf{i} - 9\mathbf{j}) + (P\mathbf{i} + 2\mathbf{j}) + (\mathbf{i} - Q\mathbf{j})$$

$$= (5 + P)\mathbf{i} + (-6 - Q)\mathbf{j}$$

$$= 0\mathbf{i} + 0\mathbf{j}$$

$\therefore$ $5 + P = 0$ and $-6 - Q = 0$ $\Rightarrow$ $P = -5$ and $Q = -6$

EXERCISE 6a

In this exercise all forces are measured in newtons.
In questions 1 to 3 the forces shown in the diagram are in equilibrium.
Find the values of P, Q and, where appropriate, the value of θ.

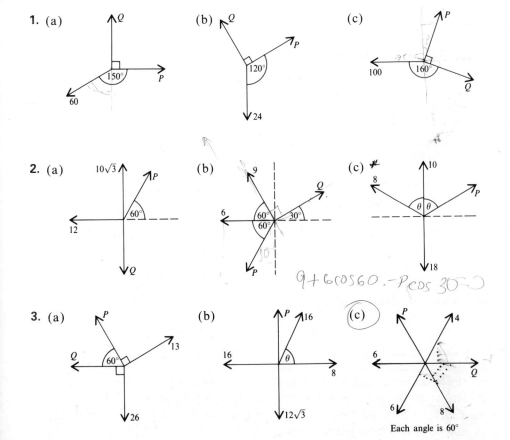

4. A light inextensible string is of length 50 cm. It is fixed to a wall at one end A and a particle of mass 4 kg is attached to the other end B. A horizontal force applied to the end B holds the particle in equilibrium at a distance of 30 cm from the wall. Find, in terms of g, the tension in the string.

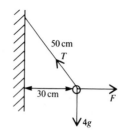

5. A small block of weight 20 N is attached to two light inelastic strings. The other ends of the strings are fixed to two fixed points on the same level, 1 m apart. The lengths of the strings are 0.6 m and 0.8 m. What is the angle between the strings? By resolving in the directions of the strings, find the tension in each string.

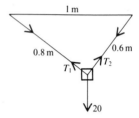

6. A small block of weight W rests on a smooth plane inclined at $30°$ to the horizontal and is held in equilibrium by a light string inclined at $30°$ to the plane. Find, in terms of W, the tension in the string.

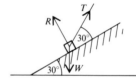

7.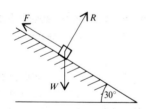

A particle of weight W rests in equilibrium on a rough plane inclined at $30°$ to the horizontal. Find, in terms of W, the magnitude of the frictional force.

8. A small block of weight $6W$ rests on a smooth plane inclined at an angle θ to the horizontal. Find θ if the block is held in equilibrium by

   (a) a force $3W$ parallel to the plane   (b) a horizontal force $2W$.

   Remember that $\tan \theta = \frac{\sin \theta}{\cos \theta}$.

9. Write down, in the form $a\mathbf{i} + b\mathbf{j}$, each of the forces shown in the diagram. Given that the forces are in equilibrium, find P and Q.

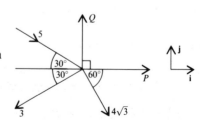

10. Four forces act on a particle keeping it in equilibrium. Find the values of p and q if the forces are

(a) $2\mathbf{i} + 7\mathbf{j}$, $5\mathbf{i} - p\mathbf{j}$, $9\mathbf{i} + 4\mathbf{j}$ and $q\mathbf{i} - 11\mathbf{j}$

(b) $\mathbf{i} - 6\mathbf{j}$, $-8\mathbf{i} + 3\mathbf{j}$, $p\mathbf{i} + q\mathbf{j}$ and $3\mathbf{i} + 10\mathbf{j}$

11. The resultant of the forces $7\mathbf{i} - 2\mathbf{j}$, $-6\mathbf{i} + 5\mathbf{j}$, $3\mathbf{i} + 6\mathbf{j}$ and $a\mathbf{i} + b\mathbf{j}$, is $11\mathbf{i} - 2\mathbf{j}$.

(a) Find the values of a and b.

(b) When a fifth force is added to the given forces, equilibrium is established. Write down in terms of $\mathbf{i}$ and $\mathbf{j}$ the force that is added.

FRICTION

Frictional forces were mentioned briefly in Chapter 4 when we considered a book lying on a table top.

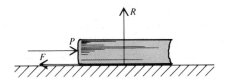

A force P applied to the book does not necessarily move it. That is because if the contact between the book and the table is rough there is frictional resistance to motion.

The frictional force F, acting on the book, acts along the table in a direction opposite to the potential direction of motion. (We know from Newton's Third Law that an equal and opposite frictional force acts on the table but this is ignored as the table is regarded as fixed.)

If P and F are the only forces acting horizontally on the book then, as long as it is stationary, P and F must be equal. (It is obvious that at no stage can we say $F > P$; if this were the case the book would move towards the pushing force!). So the amount of friction is just sufficient to prevent motion.

Now if P gradually increases, but still failing to move the book, eventually the book will be *just on the point of moving*. A further increase in the value of P will make the book move.

When the book is on the point of moving, friction is said to be *limiting*; F has reached its maximum value and the book is in *limiting equilibrium*.

Beyond this point the book moves, i.e. $P > F$.

The Coefficient of Friction

For two particular surfaces in rough contact, it can be shown experimentally that the limiting value of the frictional force is a fixed fraction of the normal reaction between the surfaces.

This fraction is called the *coefficient of friction* and it is denoted by the Greek letter μ (pronounced mew, in English), i.e. for limiting friction

$$F = \mu R$$

As this is the maximum value of the frictional force, F can take any value from zero up to μR, i.e.

$$0 \leqslant F \leqslant \mu R$$

Once an object begins to move, the frictional force opposing motion remains at the constant value μR. The *marginal* difference between the value of μ when friction is limiting (the coefficient of static friction) and its value once motion takes place (the coefficient of dynamic friction) is so small that at this level it can be ignored.

The value of μ depends upon the materials of which the *two* surfaces in contact are made – it is *not* a property of *one* surface – so ideally we should always refer to *rough contact* rather than to a rough plane, etc.

However, because wording of the strictly correct definition is a bit lengthy, it is not always used and a phrase such as 'a ladder rests against a rough wall' is often found. This should be taken to mean that there is friction between the ladder and the wall. Similarly 'a block moves on a smooth plane' means that we ignore friction between the block and the plane.

It is interesting to note that rough contact does not necessarily involve surfaces that would ordinarily be described as rough. For instance, a highly polished metal block placed on a highly polished flat metal sheet is extremely difficult to move across the sheet, although each surface on its own would be called smooth. Clearly in the context of mechanics the ordinary meanings of 'rough' and 'smooth' cannot be used; instead
    'rough' means that there *is* friction at the contact;
    'smooth' means that we ignore friction at the contact.

The properties of friction discussed so far can be summarised to give what are known as the *laws of friction*.

3. A particle of weight 8 N is resting in rough contact with a plane inclined at an angle α to the horizontal where $\tan \alpha = \frac{3}{4}$. The coefficient of friction between the particle and the plane is μ. A horizontal force P newtons is applied to the particle. When $P = 16$ the particle is on the point of slipping up the plane.

(a) Find μ.

(b) Find the value of P such that the particle is just prevented from slipping down the plane.

(a)

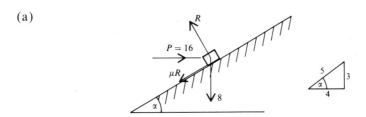

Resolving parallel and perpendicular to the plane involves μ in only one equation.

Resolving $\nearrow$ $16 \cos \alpha - \mu R - 8 \sin \alpha = 0$

$\Rightarrow$ $16 \times \frac{4}{5} - \mu R - 8 \times \frac{3}{5} = 0$

$\Rightarrow$ $\mu R = 8$

Resolving $\nwarrow$ $R - 16 \sin \alpha - 8 \cos \alpha = 0$

$\Rightarrow$ $R - 16 \times \frac{3}{5} - 8 \times \frac{4}{5} = 0$

$\Rightarrow$ $R = 16$

$\mu R / R = \frac{8}{16}$ $\Rightarrow$ $\mu = \frac{1}{2}$

(b)

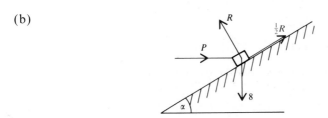

This time resolving horizontally and vertically uses P in only one equation.

Resolving $\uparrow$ $R \cos \alpha + \frac{1}{2} R \sin \alpha - 8 = 0$

$\Rightarrow$ $R = \frac{80}{11}$

Resolving $\rightarrow$ $P + \frac{1}{2} R \cos \alpha - R \sin \alpha = 0$

$\Rightarrow$ $P = \frac{R}{5} = \frac{16}{11}$

i.e. $P = 1\frac{5}{11}$

4. The diagram shows a plane, inclined at an angle α to the horizontal, where $\tan \alpha = \frac{3}{4}$, with a smooth pulley fixed at the top. A light inextensible string passes over the pulley and connects the particle A, which is in rough contact with the plane, to the particle B hanging freely. The mass of B is 3 kg, the mass of A is 2 kg and the coefficient of friction with the plane is 0.2. When the system is released from rest find the acceleration of the particles and the tension in the string. Take the value of g as 10 and give answers corrected to 2 significant figures.

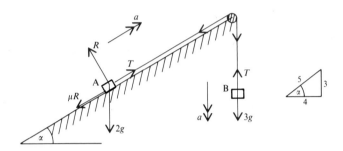

When the system moves, the frictional force has a constant value of $0.2R$.

Using Newton's Law, $F = ma$, in the direction of motion of each particle gives

For B $\downarrow$ $3g - T = 3a$ [1]

For A $\nearrow$ $T - 2g \sin \alpha - 0.2R = 2a$ [2]

            $\nwarrow$ $R - 2g \cos \alpha = 0$ [3]

From [3] $R = 20 \times \frac{4}{5} = 16$

In [2] $T = 2a + 20 \times \frac{3}{5} + 0.2 \times 16 = 2a + 15.2$

In [1] $3a = 30 - (2a + 15.2) \qquad \Rightarrow \qquad 5a = 14.8$

$\therefore$ $a = 2.96$ and $T = 21.12$

To 2 sf, the acceleration is 3.0 m/s^2 and the tension is 21 N.

EXERCISE 6b

In each question from 1 to 8 the particle is of weight 24 N and has rough contact with the specified surface; μ is the coefficient of friction between the particle and the surface. Give answers that are not exact corrected to 3 significant figures.

1. The particle is just about to slip down a plane inclined at 30° to the horizontal. Find the value of μ.

2. The particle is on a horizontal plane and is being
pulled by a horizontal string. If it is just on the
point of moving when the tension in the string is
8 N, find the value of μ.

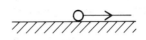

3. The particle is just about to slip up a plane inclined at 30° to the horizontal,
when being pushed by a force parallel to the plane. If $\mu = \frac{1}{2}$, find the
magnitude of the force.

4. The particle is on a horizontal plane and is being pulled by string inclined at
60° to the horizontal. If it is just on the point of moving when the tension in the
string is 16 N, find the value of μ.

5. The particle is supported in limiting
equilibrium on a plane inclined at 30° to the
horizontal, by a string parallel to the plane.
If $\mu = \frac{1}{5}$, find the tension in the string.

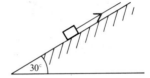

6.

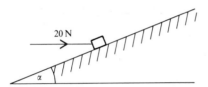

The particle is resting on a plane inclined at an angle α to the horizontal, where
$\tan \alpha = \frac{5}{12}$. A horizontal force of 20 N is on the point of moving the particle up
the plane. Find the value of μ.

7. The particle is resting on a plane inclined at an angle α to the horizontal,
where $\tan \alpha = \frac{4}{3}$. A force of 12 N parallel to the plane is just able to prevent
the particle from slipping down the plane. Find the value of μ.

8. The particle is held in limiting equilibrium, on a
plane inclined at 30° to the horizontal, by a string
inclined at 30° to the plane as shown. Given that
the value of μ is $\frac{1}{4}$, find the tension in the string
when the particle is on the point of moving

(a) up the plane

(b) down the plane.

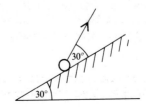

9. A small block of weight W is placed on a plane inclined at an angle θ to the horizontal. The coefficient of friction between the block and the plane is μ.

   (a) When $\theta = 30°$ the block is on the point of slipping. Show that $\mu = \frac{\sqrt{3}}{3}$.

   (b) If $\mu = \frac{1}{5}$ and $\tan\theta = \frac{3}{4}$ find, in terms of W, the magnitude of the horizontal force needed to prevent the block from slipping down.

   (c) If $\mu = \frac{2}{5}$ and $\tan\theta = \sqrt{3}$ find, in terms of W, the magnitude of the horizontal force that will be on the point of making the block slide up the plane.

10. A horizontal force P newtons is applied to a body of weight 80 N, standing in rough contact with a horizontal plane. The coefficient of friction between the body and the plane is $\frac{1}{2}$. What is the magnitude of the frictional force when

    (a) $P = 10$ (b) $P = 40$ (c) $P = 50$?

    State in each case whether or not the body moves.

11. The diagram shows a particle A lying in rough contact with a table. A light inelastic string attached to A, passes over a smooth pulley at the edge of the table and is attached to another particle B hanging freely. The particles are of equal mass M and the coefficient of friction between A and the table is $\frac{2}{5}$. Find in terms of g and M, the acceleration of B and the tension in the string.

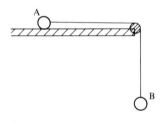

12.

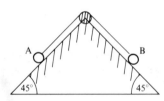

A and B are two particles connected by a light string that passes over a smooth pulley at the top of a wedge as shown in the diagram. The mass of A is m and that of B is $2m$. Contact between each particle and the wedge is rough with a coefficient of friction of $\frac{1}{5}$. When the system is allowed to move, find the acceleration and the tension in the string.

CHAPTER 7

WORK AND POWER

WORK

Anyone pushing a heavy crate across a storeroom floor would be justified in thinking that it was hard work. This is a common concept of work, i.e. making an effort to move an object, and it is reflected in the following definition of *mechanical work*.

When an object moves under the action of a constant force F, the amount of work done by the force is given by:
   the component of F in the direction of motion × distance moved.

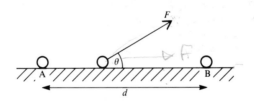

So if a constant force F moves an object from A to B,
the amount of work done by F is $(F \cos \theta) \times (d)$,
i.e. $Fd \cos \theta$

If the force is measured in newtons and the distance in metres, the work done is measured in joules (J).

For example, if a force of 12 N acts on a body and moves it a distance of 3 m in the direction of the force, the amount of work done by the force is 36 J.

Note that work is done *only if the force succeeds in moving the object*.
A force applied to an object that remains at rest, does not do any work.

When several forces act on one body, the work done by each force can be found independently of the others.

Work Done Against a Particular Force

In some circumstances we are more concerned about the work that is needed to *overcome* an opposing force. Consider, for example, an object that is pulled at constant speed from A to B, a distance *s* along a rough surface.

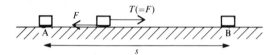

A frictional force of magnitude *F* acts on the object, opposing the motion.

Because there is no acceleration, the force causing the displacement is equal and opposite to the frictional force, i.e. it is of magnitude *F* and therefore the amount of work it does is given by *Fs*.

This work is done to overcome friction and is called *the work done against friction*. So we see that

the work done *against* a force is given by:
the magnitude of that force × the distance moved in the opposite direction.

Now consider a body of mass *m*, raised vertically through a height *h*.
The weight, *mg*, acts vertically downward and has to be overcome by an upward force in order to raise the body.

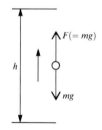

The force needed to raise the body vertically at constant speed is *mg* upwards. The work done by this force is *mg* × *h*. This amount of work is needed to overcome the opposing force of gravity, i.e. the work done against gravity is *mgh*.

Again we see that the work done against gravity is given by

the magnitude of the gravitational force (*mg* downwards)
× the distance moved in the opposite direction (*h* upwards).

It is sometimes convenient to regard work done against a force as being negative work done *by* that force. In the situation above, for example, the work done *by* gravity is $-mgh$.

Work Done by a Moving Vehicle

A variety of forces can act on a moving vehicle, including friction, air resistance, the weight of the vehicle, reaction with the ground etc., but most important is the driving force.

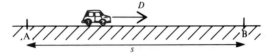

The *work done by a vehicle*   means   the *work done by the driving force*,
i.e. *D* × *s*.

Examples 7a

Whenever a force is represented on a diagram by a letter, e.g. *P*, it is understood that the force is *P* newtons.

1. A body resting in smooth contact with a horizontal plane, moves 2.6 m along the plane under the action of a force of 20 N. Find the work done by the force if it is applied

   (a)   horizontally

   (b)   at 60° to the plane.

   (a)

   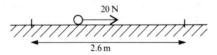

   The whole of the force acts in the direction of motion.

   Work done = 20 × 2.6 J = 52 J

   (b)

   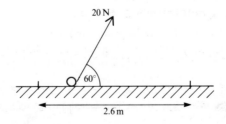

   The component of the force that acts in the direction of motion is *F* cos 60°

   Work done = (20 cos 60° × 2.6) J = (20 × $\frac{1}{2}$ × 2.6) J = 26 J

2. Sixteen crates, each of mass 250 kg, are raised 3 m by a hoist, to be placed on a platform. Find the work done against gravity by the hoist.

The work done against gravity in raising one crate $= (250 \times g \times 3)$ J

$\qquad\qquad\qquad\qquad\qquad\qquad\qquad\qquad\qquad\quad = 750g$ J

The work done against gravity in raising 16 crates $= (16 \times 750g)$ J

$\qquad\qquad\qquad\qquad\qquad\qquad\qquad\qquad\qquad\qquad\quad = 12\,000g$ J $= 12g$ kJ

(1000 J is 1 kilojoule, i.e. 1 kJ)

3. A small block of mass 2 kg slides, at constant speed, $\frac{4}{5}$ m down the face of a plane inclined at $30°$ to the horizontal. Contact between the block and the plane is rough. Giving answers in terms of g, find

(a)   the work done by
     (i)   the weight of the block
    (ii)   the reaction between the block and the plane.

(b)   the work done against friction.

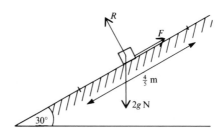

(a)   The direction of motion is down the plane.
     (i)   The component of weight down the plane is $2g \sin 30°$ N $= g$ N
        $\therefore$ the work done by the weight is $g \times \frac{4}{5}$ J $= \frac{4}{5}g$ J
    (ii)   The reaction has no component parallel to the plane.
        $\therefore$ no work is done by the reaction.

(b)   The acceleration of the block is zero so the resultant force along the plane is zero.

Resolving $\nearrow$ gives $2g \sin 30° - F = 0$

$\therefore$ the frictional force is g N up the plane.

$\therefore$ the work done against friction is $(g \times \frac{4}{5})$ J $= \frac{4}{5}g$ J

4. A tractor climbs, at a steady speed of 5 m/s, up a slope inclined at an angle α to the horizontal. If the mass of the tractor is 1400 kg and $\sin \alpha = \frac{3}{25}$, find the work done by the tractor against gravity per minute. If the total work done by the tractor in this time is 780 kJ, find the resistance to motion. (Take g as 10.)

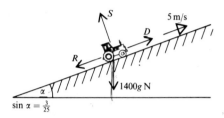

In one minute:

    the tractor climbs up the slope a distance 5×60 m $= 300$ m

$\therefore$ the vertical distance climbed is $300 \sin \alpha$ m $= 36$ m

$\therefore$ the work done against gravity is $(1400 \times g \times 36)$ J $= 504$ kJ

The speed of the tractor is constant so the acceleration is zero.

Resolving $\nearrow$ gives $D - R - mg \sin \alpha = 0$

$\therefore$ $D = R + 1400 \times 10 \times \frac{3}{25} = R + 1680$

The work done by the tractor $=$ the work done by the driving force

                                $= (D \times 300)$ J

                                $= (R + 1680) \times 300$ J

This is known to be 780 kJ

Therefore $780\,000 = 300(R + 1680)$ $\Rightarrow$ $R = 2600 - 1680$

The resistance to motion is 920 N.

EXERCISE 7a

In each question from 1 to 3, a small object moves from A to B under the action of the forces shown in the diagram. Find the work done by each force.

1. AB $= 2$ m

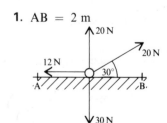

2. AB $= 3$ m

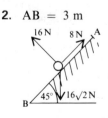

3. AB $= 4$ m

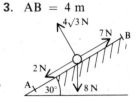

For the rest of this exercise use $g = 10$ and give answers corrected to
2 significant figures.
In each question from 4 to 8 find the work done against gravity.

4. A block of mass 3 kg is raised vertically through 2.1 m.

5. A workman of mass 87 kg climbs a vertical ladder of length 7 m.

6. Eight crates of beer, each of mass 24 kg, are lifted from the ground on to a
shelf that is 1.8 m high.

7. A forklift truck loads a tiger in a cage into the hold of an aircraft. The
combined mass of tiger and cage is 340 kg and the floor of the hold is 7.2 m
above the ground.

8. A crane lifts a one-tonne block of stone out of a quarry that is 11 m deep.

9.

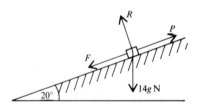

A block of mass 14 kg is pulled a distance of 6 m up a plane inclined at 20° to
the horizontal. The contact is rough and the magnitude of the frictional force
is 30 N.

Find the work done against (a) friction (b) gravity.

10. There is an average resistance of 480 N to the motion of a train as it travels
5.8 km between two stations. Find the work done against the resistance.

11.

 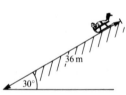

A boy, whose mass is 40 kg, has a sledge of mass 6 kg. He pulls the sledge 36 m
up a slope inclined at 30° to the horizontal.

(a) Find the amount of work done by the boy against gravity in pulling the
sledge up.

The boy then sits on the sledge and slides back to the foot of the slope.

(b) Find the work done by gravity during the descent.

12. A wardrobe is lowered by a rope from the balcony of a fifth-floor flat to the ground, 12 m below. Given that the mass of the wardrobe is 37 kg, find the work done by the rope during the descent.

In questions 13 to 16 a box of mass 6 kg is pulled by a rope along a horizontal surface at a constant speed.

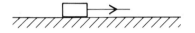

13. The speed is 4 m/s. Find the work done by the rope in 20 seconds if the tension in it is 18 N.

14. The work done by the rope in moving the box 8 m is 200 J. Find the tension in the rope.

15. The coefficient of friction between the box and the surface is $\frac{1}{3}$. If friction is the only resistance to the motion of the box, find the work done by the rope in pulling the box through 5 m.

16. The work done by the rope against friction in pulling the box a distance of 12 m is 180 J. Find the coefficient of friction between the box and the surface.

17.

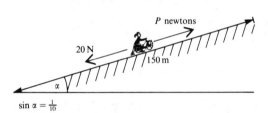

A girl pushes her bicycle 150 m up a hill inclined at an angle α to the horizontal where $\sin \alpha = \frac{1}{10}$. If the combined weight of the girl and her bicycle is 700 N, find the work she does against gravity. If there is an average resistance to motion of 20 N find the total work done by the girl.

18. A crate of mass 40 kg is pulled by a rope at a constant 1.5 m/s down a slope inclined at 15° to the horizontal. Contact is rough and the coefficient of friction is 0.7. Find

   (a) the frictional force

   (b) the tension in the rope

   (c) the work done by the rope per second

   (d) the work done by gravity while the crate moves down the slope for 6 seconds.

19.

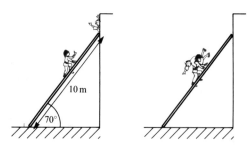

To rescue a woman trapped in a burning flat, a fireman climbs 10 m up the turntable ladder, which is inclined at an angle of 70° to the horizontal. How much work does the fireman, whose mass is 80 kg, do against gravity during his climb? He lifts the woman out through the window and carries her down the ladder to safety. If she has a mass of 50 kg,

(a)   write down the magnitude of

    (i)   the force exerted by the woman on the fireman

    (ii)   the force exerted by the fireman on the ladder

(b)   the work done by gravity during the descent.

20.

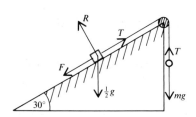

A small smooth pulley is fixed at the top of the rough face of a wedge which is inclined at 30° to the horizontal. A block A of mass $\frac{1}{2}$ kg, lying on the face, is attached to one end of a light inextensible string which passes over the pulley and carries a particle B hanging freely at the other end. The coefficient of friction between the block and the wedge is $\frac{1}{\sqrt{3}}$. If the particle B is moving down with a constant speed, find, in terms of g,

(a)   the frictional force acting on the block

(b)   the tension in the string

(c)   the weight of the particle.

When the particle moves down through 1 m, find

(d)   the work done by gravity on the particle

(e)   the total work done against gravity and friction by the string attached to the block.

POWER

There are many situations where it is not sufficient to know *how much* work a force can do. It is also important to know *the rate at which the work is being done*. This quantity is known as *power*.

One unit of power is produced when work is done at the rate of 1 joule per second. This unit is called the watt (W).

A machine working at the rate of 1 joule per second has a power of 1 watt. If 1000 joules of work are done per second, the power is 1000 watts or 1 kilowatt (1 kw).

If we know the total work done in a certain time, the *average power* can be found.

For example, a force that does 45 joules of work in 9 seconds is working at an average rate of 5 joules per second,

i.e. the average power of the force is 5 watts.

The Power of a Moving Vehicle

The power of a vehicle is defined as
the rate at which the driving force is working.

A vehicle that has a speed of v m/s is moved v metres in 1 second by the driving force, D newtons.
Therefore the work done in 1 second by the driving force of the vehicle is Dv joules.
i.e. the power of the vehicle is Dv joules/second which is Dv watts.

Therefore, if H watts is the power of a vehicle,

$$H = Dv$$

If the speed of the vehicle is constant, both D and v are constant and therefore the power is constant.
If the speed is not constant, the value of Dv gives the power *at the instant* when the speed is v m/s.

Note that if the vehicle is stationary, its power is zero. This emphasises the difference between the meaning of power in mechanics and the way the word is used in the motor trade.

There is a maximum value of the power a particular vehicle can generate. When the maximum power is used in a given situation, the speed produced is also maximum. In this condition no acceleration is possible so the resultant force acting on the vehicle is zero.

A vehicle can use less power than the maximum available if, for example, a lower speed is desirable or the resistance falls.

When solving problems involving the power of a moving vehicle, it is often helpful to express the driving force in the form $\dfrac{H}{v}$. Doing this can reduce the length of the solution.

Examples 7b

1. A bricklayer's mate can carry a hod of bricks up a vertical five-metre ladder in 46 seconds. Find the average power required if the combined mass of the man and his bricks is 92 kg. (Take g as 9.8.)

The work done against gravity is $92g \times 5 \text{ J} = 4508 \text{ J}$

This work is done in 46 s

$\therefore$ the work is done at an average rate of $\frac{4508}{46}$ joules per second

i.e. the average power is 98 W.

2. On a level track a train has a maximum speed of 50 m/s. The total resistance to motion is 28 kN.

(a) Find the maximum power of the engine.

The resistance is reduced and it is found that the power needed to maintain the same speed as before is 1250 kW.

(b) Find the lower resistance.

(a)

At maximum speed there is no acceleration and the resultant force in the direction of motion is zero.

Resolving $\rightarrow$ $\qquad D - 28\,000 = 0 \qquad \Rightarrow \qquad D = 28\,000$

To achieve maximum speed, maximum power is needed.

Maximum power = driving force $\times$ maximum velocity

$\qquad\qquad\qquad = 28\,000 \times 50 \text{ W}$

$\qquad\qquad\qquad = 1400 \text{ kW}$

(b)

As the velocity is constant the acceleration is zero, so the resultant force is zero.

$$H = Dv \qquad \Rightarrow \qquad 1\,250\,000 = D_1 \times 50 \qquad \Rightarrow \qquad D_1 = 25\,000$$

Resolving $\rightarrow \qquad 25\,000 - R_1 = 0$

The reduced resistance is 25 kN.

3. When a car of mass 1200 kg is driving up a hill inclined at $\propto$ to the horizontal, with the engine working at 32 kW, the maximum speed is 25 m/s. Given that $\sin \alpha = \frac{1}{16}$, find the resistance to motion. (Use $g = 10$.)

Speed is maximum so acceleration is zero and the resultant force up the hill is zero.

Driving force = power/velocity.

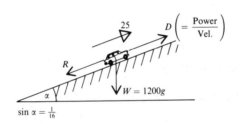

The driving force is $\dfrac{32\,000}{25}$ N = 1280 N

Resolving $\nearrow \qquad D - R - W \sin \alpha = 0$

$\Rightarrow \qquad 1280 - R - 1200g \times \frac{1}{16} = 0$

$\therefore \qquad R = 1280 - \dfrac{12\,000}{16} = 530$

The resistance to motion is 530 N.

4. The combined weight of a cyclist and his machine is 850 N. When riding, the resistive forces are proportional to the speed. On a level road, exerting his maximum power of 180 W, the maximum speed attainable is 8 m/s. Find the maximum speed which can be achieved by working at 30 W down a hill with a gradient of 7%.

(If the gradient of a hill inclined at α to the horizontal is 7%, then $\sin \alpha = 0.07$)

Resistance $\propto$ speed $\qquad \Rightarrow \qquad R = kv \qquad$ where k is a constant of proportion.

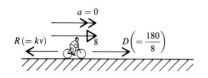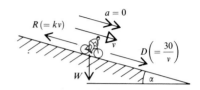

On the level

Resolving $\quad\rightarrow\quad\quad D - R = 0 \quad\quad\Rightarrow\quad\quad \dfrac{180}{8} - 8k = 0$

$$\Rightarrow\quad\quad k = \tfrac{45}{16}$$

Downhill

Resolving $\quad\searrow\quad D + W\sin\alpha - R = 0 \quad\quad\Rightarrow\quad\quad \dfrac{30}{v} + 850 \times 0.07 - \dfrac{45v}{16} = 0$

$\therefore\quad\quad 2.8125v^2 - 59.5v - 30 = 0$

Solving this quadratic equation by using the formula gives

$$v = \frac{59.5 \pm \sqrt{(59.5^2 + 4 \times 2.8125 \times 30)}}{2 \times 2.8125}$$

Only the positive solution has any significance.

Hence $\quad v = 21.64\ldots$

The maximum speed downhill is 21.6 m/s, corrected to 3 sf.

EXERCISE 7b

If the speed of a vehicle is given in km/h, the unit must be converted to m/s in order to be consistent with the other units being used. This conversion is done by using the fact that 1 km/h $= \tfrac{5}{18}$ m/s.
Use $g = 10$ and give answers corrected to 2 significant figures.

In questions 1 to 5 find the average rate at which work is done.

1. A mass of 60 kg is lifted vertically through 4 m in 9 seconds.

2. A mass of 40 kg is lifted vertically at a constant speed of 5 m/s.

3. A cat weighing 24 N climbs up a 3 metre high wall in 2 seconds.

4. A part-time assistant in a supermarket is stacking wine bottles on a shelf. Each bottle weighs 5 N and is lifted up 1.6 m. The assistant stacks 36 bottles in a minute.

5. An elevator is raising bales of straw, each with a mass of 23 kg, to the floor of a loft that is 2.1 m above ground level. On average, 64 bales are raised per hour.

6. A car driving at a constant speed v, against a constant resistance R, is working at a rate H.

(a) If $v = 25$ m/s and $R = 960$ N, find H.

(b) If $H = 60$ kW and $v = 120$ km/h, find R.

(c) If $R = 1300$ N and $H = 26$ kW, find v.

7. A goods train has a maximum speed of 90 km/h on a level track when the resistive forces amount to a constant 40 kN. Find the maximum power of the engine.

8. A car has a maximum speed of 140 km/h on a level road with the engine working at 54 kW. Find the resistance to motion.

9.

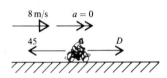
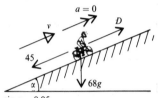

$\sin \alpha = 0.05$

A boy and his bicycle have a combined mass of 68 kg. Working at maximum power the cyclist can achieve a speed of 8 m/s on a level road against resistive forces totalling 45 N. Find the maximum power. Find also the maximum speed up an incline with a 5% gradient.

10. The maximum power that a van of mass 900 kg can exert is 36 kW. If the resistance to the motion of the van is a constant 1500 N, find the maximum speed that the van can reach on a slope of inclination 1 in 15 (i.e. $\sin \alpha = \frac{1}{15}$) when driving   (a) up the slope   (b) down the slope.

Give one reason why your answers may not reflect the actual maximum speeds of a real van.

11. A car of mass 1100 kg has a maximum power output of 44 kW. The resistive forces are constant at 1400 N. Find the maximum speed of the car

(a) on the level   (b) up an incline with gradient 5%

(c) down the same incline when using half the maximum power.

12. A car of mass 950 kg has a maximum power of 30 kW and encounters resistive forces totalling 1050 N.

(a) Find its maximum speed on level ground.

(b) The driver wishes to descend a hill of gradient 1 in 15 without increasing speed. What percentage of the maximum power of the engine should be used?

13. A lorry of mass 2000 kg is subject to a constant resistance. The maximum speed of the lorry down a slope of 1 in 10 is 24 m/s and the maximum speed up the same slope is 12 m/s. Find

(a)   the maximum power of the engine

(b)   the constant resistance

(c)   the maximum speed of the lorry on a level road.

14. The engine of a car of mass 1000 kg is working at the constant rate of 50 kW. The resistance to motion is proportional to the speed.

(a)   If the maximum speed on a level road is 40 m/s, find the constant of proportion.

(b)   Find the maximum speed up a slope with a gradient of 5%.

(c)   When the engine is working at half power, the maximum speed of the car down a slope inclined at an angle α to the horizontal is 35 m/s. Find α to the nearest degree.

ACCELERATING VEHICLES

If a vehicle exerts a driving force that exceeds all the forces opposing motion, there is a resultant force in the direction of motion. As a result, the car has an acceleration which can be found by applying Newton's Law.

Examples 7c

1. A car of mass 1200 kg has a maximum speed of 144 km/h on a level road when there is a resistive force of 56 N. Find the acceleration of the car at the instant when its speed is 81 km/h and the engine is working at maximum power.

Speeds must be expressed in metres per second.

$144 \text{ km/h} = 144 \times \frac{5}{18} \text{ m/s} = 40 \text{ m/s}$; similarly $81 \text{ km/h} = 22.5 \text{ m/s}$

At maximum speed

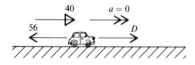

Resolving $\rightarrow$ $D - 56 = 0$ $\Rightarrow$ $D = 56$

$\therefore$ if H watts is the maximum power, $H = Dv = 56 \times 40 = 2240$

The maximum power is 2240 W.

At the lower speed

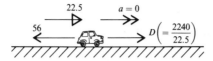

The resultant force in the direction of motion is $D - R$

Applying $F = ma$ in this direction gives

$$\frac{2240}{22.5} - 56 = 1200a \qquad \Rightarrow \qquad a = \frac{43.56}{1200} = 0.0363$$

Therefore, corrected to 2 sf, the acceleration is 0.036 m/s^2.

2. A motor cyclist whose mass combined with his machine is 240 kg is driving up
   a road of inclination 1 in 10 at maximum power of 10 kW. When the speed is
   25 m/s, the motor bike is accelerating at 0.05 m/s^2. Taking the value of g as 10,

   (a) find the constant resistance to motion.

   At the top of the hill he picks up a pillion passenger of mass 80 kg and drives on
   along the road which is now horizontal. If the resistance is increased by 20%,
   find

   (b) the greatest speed that can be achieved when the engine is working at 70%
       of the maximum power

   (c) the immediate acceleration produced if maximum power is suddenly
       engaged.

(a)

The resultant force up the hill is

$$D - W \sin \alpha - R = \frac{10\,000}{25} - 2400 \times \frac{1}{10} - R$$

Using Newton's Law, $F = ma$, in the direction of motion gives

$$400 - 240 - R = 240 \times 0.05 \qquad \Rightarrow \qquad R = 148$$

The resistance to motion is 148 N.

(b)   The resistance is now 120% of 148 N, i.e. 177.6 N, and the power is 70% of 10 kW, i.e. 7000 W.

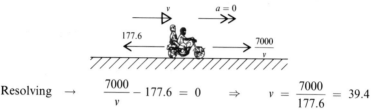

Resolving $\rightarrow$ $\dfrac{7000}{v} - 177.6 = 0$ $\Rightarrow$ $v = \dfrac{7000}{177.6} = 39.4$

The maximum speed is 39.4 m/s.

(c)

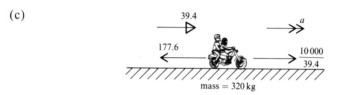

mass = 320 kg

The resultant force in the direction of motion is $D - R$

Therefore using $F = ma$ gives

$$\dfrac{10\,000}{39.4} - 177.6 = 320a \qquad \Rightarrow \qquad a = \dfrac{253.8 - 177.6}{320} = 0.24$$

The immediate acceleration is 0.24 m/s².

Note that this is an instantaneous acceleration; as a result the velocity increases, so reducing the driving force which in turn reduces the acceleration.

EXERCISE 7c

Use $g = 10$ and give answers corrected to 2 significant figures.

1. A car of mass 1500 kg is being driven up an incline of 1 in 20 against a constant resistance to motion of 1000 N.

   (a)   At the instant when the speed is 20 m/s and the acceleration is 0.1 m/s², find the power being exerted by the engine.

   (b)   If the engine is working at 20 kW, find the acceleration at the instant when the speed is 7.5 m/s.

2. The resistive forces opposing the motion of a car of mass 2000 kg total 5000 N. If the engine is working at 70 kW, find the acceleration at the instant when the speed is 40 km/h.

3. A car of mass 1000 kg has a maximum power of 50 kW. The car is travelling up a hill with a gradient of 8% against resistance to motion of 3000 N. Find the acceleration at the instant when the speed is 30 km/h.

4. A train of mass 400 tonnes is moving down an incline of 1 in 50 using a power output of 50 kW. The resistance to motion is 30 kN. Find the acceleration at the instant when the speed is 20 m/s.

5. The constant resistances to the motion of a car of mass 1200 kg total 960 N.

   (a) If the car is driving along a level road and has an acceleration of 0.2 m/s^2 at the instant when the speed is 25 m/s, find the power exerted by the engine.

   (b) If the car is moving down a slope of inclination 1 in 15 and working at 40 kW, find the acceleration at the instant when the speed is 25 m/s.

6. The engine of a train of mass 50 000 kg is working at 1800 kW as the train ascends a slope of inclination 1 in n. The train encounters constant resistive forces of 10 kN.

   (a) If the maximum speed of the train is 50 m/s, find n.

   (b) Find the acceleration at the instant when the speed is 30 m/s.

7. A car of mass 1000 kg encounters resistive forces of 1200 N when ascending a slope with a 10% gradient.

   (a) If the engine is working at 30 kW find the maximum speed.

   (b) When moving at this speed the driver suddenly increases the power of the engine by 25%. Find the immediate acceleration of the car.

8. A cyclist is riding up an incline of 1 in 20. Working at 2 kW the maximum speed up the incline is 20 km/h.

   (a) Find the resistance to motion given that the combined mass of the cyclist and machine is 100 kg.

   (b) If there is no change in resistance or power, find the instantaneous acceleration when the cyclist reaches level ground at the top of the slope.

9. A car of mass 800 kg moves against a constant resistance R newtons. The maximum speeds of the car up and down an incline of 1 in 16 are respectively 14 m/s and 42 m/s. If the rate at which the engine is working is H kW, find

   (a) the values of R and H

   (b) the acceleration at the instant when the speed is 17.5 m/s on level ground.

10. The resistance to the motion of a car of mass 1000 kg is proportional to the square of the speed. With the engine working at 60 kW, the car can drive up an incline of 1 in 20 at a steady speed of 30 m/s. If the car travels down the same slope with the engine working at 40 kW, find the acceleration at the instant when the speed is 20 m/s.

CHAPTER 8

MECHANICAL ENERGY

ENERGY

Anything that has the capacity to do work, possesses energy. This energy can be used up in doing work.

Conversely, in order to give energy to an object, work must be done to it,

i.e. work and energy are interchangeable and so are measured in the same unit, the joule.

There are various different forms of energy such as light, heat, sound, electrical energy and chemical energy. These can often be converted from one form to another, e.g. electrical energy can be used to give heat or light energy.

In this book however we are concerned primarily with *mechanical energy,* which is the capacity to do work as a result of motion or position.

KINETIC ENERGY (KE)

A body moving with speed v possesses *kinetic energy.* The value of the KE is equal to the amount of work needed to bring that body from rest to the speed v and an expression for its value can be found as follows.

Consider a body of mass m which starts from rest and reaches a speed v after moving through a distance s under the action of a constant force F.

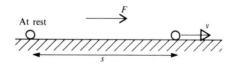

The acceleration, a, is given by

$$v^2 - u^2 = 2as \qquad \Rightarrow \qquad a = \frac{v^2 - 0}{2s}$$

Then Newton's Law, $F = ma$, gives

$$F = \frac{mv^2}{2s} \quad \Rightarrow \quad Fs = \tfrac{1}{2}mv^2$$

Now Fs is the work done in producing the kinetic energy, therefore $\tfrac{1}{2}mv^2$ is the value of the kinetic energy, i.e.

$$\textbf{KE} = \tfrac{1}{2}\boldsymbol{mv}^2$$

Note that both m and v^2 are always positive quantities showing that KE is always positive and does not depend upon the direction of motion, i.e. KE is a scalar quantity.

POTENTIAL ENERGY (PE)

Potential energy is a property of position. If a body is in such a position that, if released, it would begin to move, it possesses PE.

Consider, as an example, a body that is held at a height h above a fixed level. If that body is released it will begin to fall, i.e. it will begin to possess KE. So before it is released it has the *potential* to move, hence the name for energy due to position.

The value of the PE is equal to the work needed to raise the body through a vertical distance h.

The work done in raising a body of mass m is the work done against gravity,

i.e. $mg \times h$

      $PE = mgh$

If the body falls from rest and reaches a speed v at the bottom then, using $v^2 - u^2 = 2as$ gives $v^2 = 2gh$, i.e. $gh = \tfrac{1}{2}v^2$.

Therefore $mgh = \tfrac{1}{2}mv^2$ confirming that potential energy is converted into kinetic energy.

There is no absolute value for the PE of an object, as the height h is measured from some particular fixed level. If a different level is chosen the PE is changed without the body itself moving. It follows therefore that in every problem the level from which height is measured must be clearly specified. As the PE of an object that is *on* the chosen level is zero, in this book we identify this datum by marking it 'PE = 0'.

Negative Potential Energy

If an object is *below* the datum the value of *h* is negative (*h* is the height *above* the datum), so the object has *negative potential energy*.

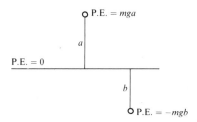

Note that there is another type of potential energy, which readers who go on to Module F will meet. It is called elastic potential energy (EPE) and it is a property of an object attached to a stretched elastic string. None of the work in this book requires knowledge of this type of mechanical energy but it is as well to be aware that it exists.

Examples 8a

1. A window cleaner of mass 72 kg climbs up a ladder to a second-floor window, 5 m above ground level. What is his potential energy relative to the ground? He then descends 3 m to clean a first-floor window. Find how much potential energy he has lost. (Use $g = 9.8$.)

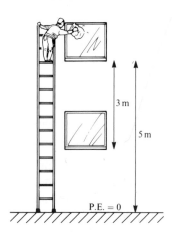

At the second-floor window, A,

$$m = 72$$

$$g = 9.8$$

$$h = 5$$

$$\therefore \quad PE = mgh = 72 \times 5 \times 9.8 \text{ J}$$

$$= 3528 \text{ J}$$

In descending to the lower window, B, the reduction in height is 3 m.

$$\text{Loss in PE} = mg \text{ (reduction in } h)$$

$$= 72 \times 9.8 \times 3 \text{ J}$$

$$= 2120 \text{ J} \text{ (2 sf)}$$

Alternatively we could find the values of PE at the two windows and subtract.

2. A particle of mass 6 kg has a velocity v metres per second.
What is its kinetic energy if   (a) $v = 7$ (b) $\mathbf{v} = 4\mathbf{i} - 3\mathbf{j}$?

(a) $KE = \frac{1}{2}mv^2 = \frac{1}{2} \times 6 \times 7^2 \text{ J} = 147 \text{ J}$

(b)   The speed of the particle is $|\mathbf{v}|$

   When $\mathbf{v} = 4\mathbf{i} - 3\mathbf{j}$, $|\mathbf{v}| = \sqrt{(4^2 + [-3]^2)} = 5$

   $KE = (\frac{1}{2} \times 6 \times 5^2) \text{ J} = 75 \text{ J}$

3. A bird of mass 0.6 kg, flying at 9 m/s, skims over the top of a tree 6.2 metres
high. What is the total mechanical energy of the bird as it clears the tree?
Use $g = 9.8$.

KE of the bird is $\frac{1}{2}mv^2$,

i.e. $\frac{1}{2} \times 0.6 \times 9^2 \text{ J} = 24.3 \text{ J}$

PE of the bird is mgh

i.e. $0.6 \times 9.8 \times 6.2 \text{ J} = 36.5 \text{ J}$

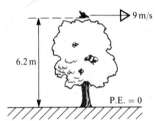

The total mechanical energy is $KE + PE = 60.8 \text{ J}$

4. Water is being raised by a pump from a storage tank 3 m below ground level
and ejected at ground level through a pipe at 6 m/s. If the water is delivered at a
rate of 420 kg each second, find the total mechanical energy supplied by the
pump in one second in lifting and ejecting the water. (Take g as 9.8.)

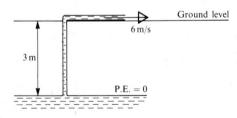

In 1 second,   PE gained by water $= mgh$

   $= 420 \times 9.8 \times 3 \text{ J} = 12\,348 \text{ J}$

   KE gained by water $= \frac{1}{2}mv^2$

   $= \frac{1}{2} \times 420 \times 6^2 \text{ J} = 7560 \text{ J}$

The total energy gained by the water is supplied by the pump.

Total ME supplied by the pump per second $= 19\,908 \text{ J}$.

EXERCISE 8a

Use $g = 9.8$ and give answers corrected to 2 or 3 significant figures as appropriate.

1. The potential energy of a particle of mass m kilograms, which is at a height h metres above a given datum, is N joules.

   (a) If $m = 4$ and $h = 11$, find N.   (b) If $N = 48$ and $m = 6$, find h.

2. Find, in joules, the kinetic energy of

   (a) a block of mass 8 kg moving at 9 m/s

   (b) a car of mass 1200 kg travelling at 36 km/h

   (c) a bullet of mass 16 g moving at 500 m/s

   (d) a body of mass 10 kg with a velocity $\mathbf{v}$ m/s where $\mathbf{v} = 3\mathbf{i} + 2\mathbf{j}$.

3. Find the items missing from the following table.

Mass	Speed	Kinetic energy
7 kg	2 m/s	
14 kg		126 J
	6 m/s	396 J

4.

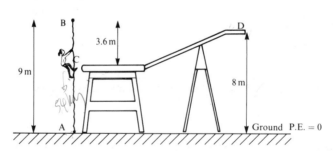

On an assault course a woman of mass 54 kg starts from ground level at A, climbs 9 m up a scramble net, drops from the top B, to a mat C, 3.6 m below B, then runs up a bar to D at a height of 8 m above the ground. Find

   (a) her potential energy relative to the ground at   (i) A   (ii) D

   (b) the gain in potential energy between A and B

   (c) the loss in potential energy between B and C

   (d) the gain in potential energy between C and D

   (e) Using the answers to parts (b), (c) and (d), find the P.E. at D.
        Check that this agrees with your answer to part (a) (ii).

5. (a) Find the gain in kinetic energy when the speed of a body of mass 4 kg increases from 7 m/s to 11 m/s.

   (b) Find the kinetic energy lost when the speed of the same body falls from 18 m/s to 5 m/s.

6. A car of mass 900 kg is travelling at 72 km/h.

(a) Find how much kinetic energy is lost if the speed falls to 54 km/h.

(b) If the kinetic energy rises to 281.25 kJ, at what speed is the car travelling?

7. A particle of mass 3 kg is at rest. It begins to move with constant acceleration and five seconds later it has kinetic energy of 150 J. Find

(a) the speed at the end of the 5 seconds

(b) how far the particle has travelled in this time.

8. A pump raises 45 kg of water through a vertical distance of 12 m.

(a) How much potential energy is gained by the water?

The water is then forced through a pipe at 12 m/s.

(b) How much kinetic energy does the water gain?

9. A pump raises water from a depth of 5 m and ejects it through a pipe with a speed of 8 m/s.

(a) If the cross-sectional area of the pipe is 0.06 m^2, find the volume of water discharged per second.

(b) Given that 1 m^3 of water has a mass of 1000 kg, find the mass of water discharged per second.

(c) Find the total mechanical energy gained by the water.

THE PRINCIPLE OF WORK AND ENERGY

Mechanical energy was defined as the capacity of a body to do work. Now the link between energy and work done can be expressed more precisely.

The work done by external forces acting on a body is equal to the increase in the mechanical energy of the body.

This is the principle of work and energy and also applies in the form:

If a body does some work, the loss in mechanical energy of the body is equal to the amount of work done.

The weight of an object is not counted as an external force in this context because work done by weight is Potential Energy and is already accounted for.

In solving problems involving this principle it is wise to use it in the form

Final ME $\sim$ Initial ME = Work Done   ($\sim$ means *the difference between*)

This avoids any confusion in cases where one type of energy increases and the other decreases.

Examples 8b

1. A force acting on a body of mass 6 kg, moving horizontally, causes the speed to increase from 3 m/s to 8 m/s. How much work is done by the force? If the magnitude of the force is 11 N, how far does the body move during this speed change?

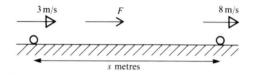

As the body is moving horizontally there is no change in PE so only KE changes.

$$\text{Work done} = \text{Final ME} - \text{Initial ME}$$
$$= \tfrac{1}{2}(6)(8^2) - \tfrac{1}{2}(6)(3^2) \text{ J}$$
$$= 165 \text{ J}$$
$$\text{Work done by force} = Fs$$
$$\therefore \qquad 165 = 11s \qquad \Rightarrow \qquad s = 15$$

The body moves 15 m.

2. A small block of mass 3 kg is moving on a horizontal plane against a constant resistance of R newtons. The speed of the block falls from 12 m/s to 7 m/s as the block moves 5 m. Find the magnitude of the resistance.

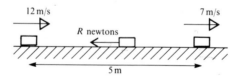

There is no change in PE and, as the speed is reducing, the final KE is less than the initial KE

$$\text{Initial ME} - \text{Final ME} = \tfrac{1}{2}(3)(12^2) - \tfrac{1}{2}(3)(7^2)$$
$$= 216 - 73.5 \text{ J}$$
$$= 142.5 \text{ J}$$
$$\text{Work done by resistance} = \text{Change in ME}$$
$$\therefore \qquad R \times 5 = 142.5 \qquad \Rightarrow \qquad R = 28.5$$

The magnitude of the resistance force is 28.5 N.

3. A stone falls vertically downward through a tank of viscose oil. The speed of the stone as it enters the oil is 2 m/s and at the bottom of the tank it is 3 m/s. Given that the oil is of depth 2.4 m, find the resistance, F newtons, that it exerts on the stone whose mass is 4 kg. Take g as 9.8.

The resistance is an opposing force so the work it does reduces the ME of the stone. Both KE and PE change.

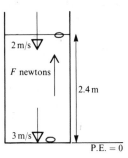

Initial ME $= [\frac{1}{2}(4)(2^2) + (4)(9.8)(2.4)]$ J

$\qquad = 8 + 94.08$ J $= 102.08$ J

Final ME $= \frac{1}{2}(4)(3^2) + 0 = 18$ J

Work done by the resistance $=$ Change in ME

$\therefore$ $F \times 2.4 = 102.08 - 18$

$\Rightarrow$ $F = 35.03\ldots$

The resistance is 35.0 N   (3 sf).

4. A car of mass 1000 kg drives up a slope of length 750 m and inclination 1 in 25. If resistance forces are negligible, calculate the driving force of the engine if the speed at the foot of the incline is 25 m/s and the speed at the top is 20 m/s. Use $g = 9.8$.

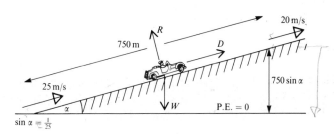

The driving force, D newtons, acts for 750 m

$\therefore$ work done by driving force $= 750D$ J

Both kinetic energy and potential energy change.

$\qquad$ Final KE $= \frac{1}{2}(1000)(20^2)$ J $= 200$ kJ

$\qquad$ Final PE $= (1000)(9.8)(750 \sin \alpha) = 9800(750 \times \frac{1}{25}) = 294$ kJ

$\therefore$ $\qquad$ Final ME $= 494$ kJ

$\qquad$ Initial KE $= \frac{1}{2}(1000)(25^2)$ J $= 312.5$ kJ

$\qquad$ Initial PE $= 0$

$\therefore$ $\qquad$ Initial ME $= 312.5$ kJ

$\qquad$ Work done by driving force $=$ Final ME $-$ Initial ME

$\therefore$ $\qquad\qquad$ $750 D = 494\,000 - 312\,500$

$\Rightarrow$ $\qquad\qquad$ $D = 242$

The driving force is 242 N.

EXERCISE 8b

Use the principle of work and energy for each question. Take g as 9.8 and give answers corrected to 2 or 3 significant figures as appropriate.

1. A mass of 6 kg is pulled by a string across a smooth horizontal plane. As the block moves through a distance of 4.2 m, the speed increases from 2 m/s to 6 m/s. Find the tension in the string.

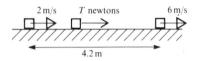

2. A body of mass 8 kg, travelling on a rough horizontal plane at 12 m/s, is brought to rest by friction. Find the work done by the frictional force.

3. A ball of mass 0.4 kg is thrown vertically upwards with a speed of 10 m/s. It comes instantaneously to rest at a height of 3.6 m above the point of projection, P. Find the resistance to its motion. The ball then falls back to P. If the resistance is unchanged, find the speed at P.

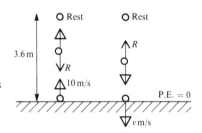

4. A body of mass 0.5 kg is lifted, by vertical force, from rest at a point A to a point B that is 1.7 m vertically above A. If the body has a speed of 3 m/s when it reaches B find the work done by the force. Hence find the magnitude of the force.

5.

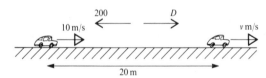

A car of mass 750 kg, is travelling along a level road at 10 m/s against a constant resistance of 200 N. Exerting a driving force of 1200 N, the driver accelerates for 20 m. At what speed will the car then be moving?

6. A block of mass 5 kg lies in contact with a horizontal plane. It is pulled from rest through a distance of 8 m by a horizontal force of 12 N. Find the speed attained if the contact between block and plane is

(a) smooth   (b) rough, with a coefficient of friction of $\frac{1}{10}$.

7. A bullet of mass 0.02 kg is fired horizontally at a speed of 360 m/s into a fixed block of wood. The bullet is embedded 0.06 m into the block. Find the average resisting force exerted by the wood.

8. A particle of mass 7 kg is pulled by a force of F newtons, 4 metres up a smooth plane inclined at $30°$ to the horizontal. Find the work done by the pulling force if

   (a) the particle is pulled at a constant speed

   (b) the speed changes from 1 m/s initially to 2 m/s at the end.

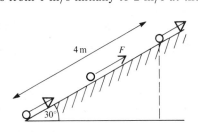

9. A block of mass 3 kg slides down a plane inclined at α to the horizontal where $\sin \alpha = \frac{1}{4}$. The block starts from rest and there is a constant frictional force of 4 N.

   (a) How far has the block travelled when its speed reaches 6 m/s?

   (b) When the block has moved 4 m down the plane, what is its speed?

10. In raising a body of mass 2 kg from rest vertically upwards by 6 m and giving it a speed of v m/s, a force does work amounting to 800 J. Find v.

11. A body of mass 2 kg falls vertically from rest through a distance of 5 m. If the speed by then is 9 m/s find the air resistance.

In questions 12 to 14, water is pumped at a rate of p m^3/s from a tank d m below ground. The water is then delivered at ground level at v m/s, through a pipe whose cross-sectional area is a m^2.
The density of water is 1000 kg/m^3.

12. If $d = 8$, $v = 10$ and $a = 0.05$, find

   (a) the volume of water discharged per second

   (b) the weight of this water

   (c) the work done per second by the pump in raising and delivering the water.

13. If $p = 0.2$, $d = 5$ and $a = 0.1$, find

   (a) the value of v

   (b) the power the pump exerts in moving the water.

14. The work done per second by the pump in raising and ejecting the water is 160 kJ. Given that $p = 6$ and $d = 2$ find v.

CONSERVATION OF MECHANICAL ENERGY

We know from the principle of work and energy that the total change in the mechanical energy of a body is equal to the work done on, or by, the body. It follows directly that:

If the total work done by the external forces acting on a body is zero there is no change in the total mechanical energy of the body, i.e. energy is conserved

This is the *principle of conservation of mechanical energy*.

Remember that the weight of a body is not an external force in this context as the work done by the weight is already included as potential energy.

At present we are concerned with only two types of mechanical energy, so a problem can be solved by working out the loss in KE, say, and equating it to the gain in PE. (However, for those readers intending to carry on studying mechanics and who will meet problems which also include the third type of mechanical energy (see p. 154) we recommend the method of equating the *total* mechanical energy in two positions.)

Examples 8c

1. A particle is projected vertically with speed 8 m/s. Find its speed after it has moved a vertical distance of 2 m   (a) upwards   (b) downwards.

Take g as 9.8 and give answers corrected to 2 significant figures.

Let the mass of the particle be m kg and the final speed be v m/s.

At A   PE $= 0$ and   ME $= \frac{1}{2}m \times 8^2$ J

$\therefore$ Total ME $= 32m$ J

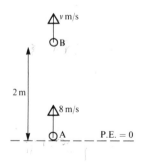

(a) At B   PE $= mgh = 19.6m$ J

and   KE $= \frac{1}{2}mv^2$ J

$\therefore$ Total ME $= (19.6m + \frac{1}{2}v^2m)$ J

Total ME at A $=$ Total ME at B

$\therefore$ $32m = 19.6m + \frac{1}{2}v^2m$

$\Rightarrow$ $v^2 = 24.8$

$\Rightarrow$ $v = 5.0$ (2 sf)

The speed at B is 5.0 m/s (2 sf)

(b) At C $PE = mg(-2)\,J$ $-19.6m$ J

    and $KE = \frac{1}{2}mv^2\,J$

     $\therefore$ Total ME $= (\frac{1}{2}mv^2 - 19.6m)\,J$

Total ME at A $=$ Total ME at C

     $\therefore$ $32m = \frac{1}{2}mv^2 - 19.6m$

     $\Rightarrow$ $v^2 = 103.2$

     $\Rightarrow$ $v = 10\ (2\ \text{sf})$

The speed at C is 10 m/s.

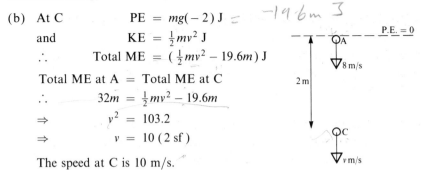

Note that in the example above the value of the mass, m, was not given and was not needed as it cancelled in each conservation of energy equation. This will always be the case as m is a factor of both PE and KE.

Note also that an alternative solution could be given using Newton's Law and the equations of motion with constant acceleration.

2. A bead is threaded on to a circular ring of radius 0.5 m and centre O, which is fixed in a vertical plane. The bead is projected from the lowest point of the ring, A, with a speed of 4 m/s, and first comes to instantaneous rest at a point B. Contact between the ring and the bead is smooth and there is no other resistance to motion. Find the height of B above A. Take g as 9.8 and give answers corrected to 2 significant figures.

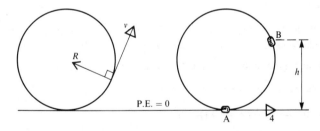

The normal reaction R is always perpendicular to the direction of motion of the bead therefore no work is done by R. No other external force is acting so energy is conserved.

At A $PE = 0$ and $KE = \frac{1}{2}mv^2 = \frac{1}{2}m(4)^2\,J$

    $\therefore$ Total ME $= (0 + 8m)\,J$

At B $PE = mgh = 9.8mh\,J$ and $KE = 0$

    $\therefore$ Total ME $= (9.8mh + 0)\,J$

       Total ME at A $=$ Total ME at B

    $\therefore$ $8m = 9.8mh$

    $\Rightarrow$ $h = \frac{8}{9.8}$

Corrected to 2 sf, B is 0.82 m above A.

3. A small block A of mass $2m$, is lying in smooth contact with a table top. A light inextensible string of length 1 m is attached at one end to A, passes over a smooth pulley at the edge of the table, and carries a block of mass m hanging freely at the other end. Initially A is held at rest, 0.8 m from the edge of the table. If the system is released, find A's speed when it reaches the edge. Take g as 9.8 and correct your answers to 2 significant figures.

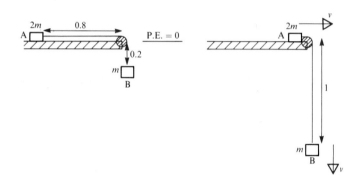

Initially   KE $= 0$,   PE of A $= 0$,   PE of B $= mg(-0.2)$ J

Total ME $= -(0.2)(9.8)m$ J $= -1.96m$ J

When A reaches the edge of the table, the speed of each block is v m/s and B is 1 m below the table top.

When A reaches the edge,

             KE of A $= \frac{1}{2}(2m)v^2$ J   KE of B $= \frac{1}{2}mv^2$ J

             PE of A $= 0$ PE of B $= mg(-1)$ J

Total ME $= (mv^2 + \frac{1}{2}mv^2 - mg)$ J

Initial total ME $=$ Final total ME

$\therefore$ $-1.96m = mv^2 + \frac{1}{2}mv^2 - 9.8m$

i.e. $7.84 = \frac{3}{2}v^2$ $\Rightarrow$ $v^2 = 5.227$ $\Rightarrow$ $v = 2.3$ (2 sf)

The speed of each block is 2.3 m/s.

Remember that the choice of the datum level for PE is arbitrary. We could just as well have chosen the lowest level reached by B. (You may like to check that the same result would be obtained.)

In Examples 2 and 3 you will notice that the letter m is used both for mass and metre. It is easy to see the difference between the two m's in type because, for mass, m is italic. When hand-writing a solution, however, this cannot be done so it is a good idea to write out the word metre in full.

EXERCISE 8c

For each question use the Conservation of Mechanical Energy. Take g as 9.8 and give answers corrected to 2 or 3 significant figures as appropriate.

1. A particle of mass m kilograms is projected vertically upwards with speed 6 m/s. Find the height it attains before first coming to rest.

2. A stone is thrown vertically downwards with speed 3 m/s. Find its speed after it has fallen 4.5 m.

3. A ball is thrown vertically upwards, from a point A, with speed 8 m/s. Given that A is 1 m above ground, find

(a) the greatest height reached by the ball
(b) the speed of the ball when its height is 2.4 m
(c) the speed of the ball as it hits the ground.

Questions 4 to 6 concern a particle P, moving on a plane inclined to the horizontal at 30°. A and B are two points on the plane. Contact between the body and the plane is smooth.

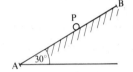

4. P is projected up the plane from A with speed 4 m/s and comes to instantaneous rest at B. Find the distance AB.

5. P is released from rest at B. Find its speed as it passes through A if AB = 2.4 m.

6. P is moving down the plane and passes through B with speed v m/s. If AB = 2.1 m and the particle passes through A with speed 6 m/s, find v.

7.

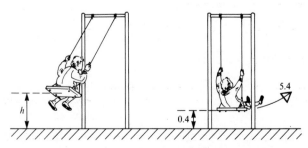

The seat of a swing is 0.4 m above the ground when it is stationary. A girl is swinging so that she passes through the lowest point with speed 5.4 m/s. Find the height of the seat above ground when she first comes to rest.

In questions 8 to 10 a smooth bead is threaded on to a smooth circular wire with centre O and radius a metres. The wire is fixed in a vertical plane.

8. The bead is released from rest at a point level with O. If $a = 0.5$, find the speed of the bead as it passes through the lowest point.

9. The bead is projected from the lowest point on the wire with speed 4.2 m/s. If $a = 0.6$, find the height above O at which the bead first comes to rest.

10. The bead is projected from the lowest point and just reaches the highest point. Given that $a = 0.8$, find the speed of projection.

For questions 11 and 12, give answers in terms of ℓ and g.

11. Two identical particles of mass m are connected by a light inelastic string of length 2ℓ. One particle, A, rests in smooth contact with a horizontal table and the other particle B hangs freely over the edge of the table. The string is perpendicular to that edge. If A is released from rest when it is at a distance ℓ from the table edge, find its speed when it reaches the edge.

12. Two particles P and Q are connected by a light inextensible string that passes over a smooth pulley. The masses of P and Q are m and $2m$ respectively. The particles are released from rest when each is at a depth 2ℓ below the pulley. Find their speed when each has moved a distance ℓ.

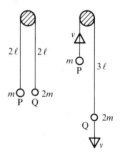

CONSOLIDATION B

SUMMARY

Newton's Laws of Motion

1. A body's state of rest or constant velocity is unchanged unless a force acts on it.

2. The acceleration of a body is proportional to the force producing it, i.e. $F = ma$, where the unit of force, called the newton, is the force needed to give a mass of 1 kg an acceleration of 1 m/s^2.
The force and the acceleration are in the same direction.

3. Forces between objects act in equal and opposite pairs.

Weight

Weight is the force of gravity attracting an object to the earth. The acceleration g which it produces is approximately 9.8 m/s^2.

The weight of an object of mass m is mg.

Equilibrium

A body that is at rest or moving with constant velocity is in a state of equilibrium and so is the set of forces acting on that body.

A particle is in equilibrium if the resultant force acting on it is zero.

Friction

Friction exists if two objects are in rough contact and have a tendency to move.

The frictional force F is just large enough to prevent motion, up to a limiting value.

When the limiting value is reached, $F = \mu R$ where R is the normal reaction and μ is the coefficient of friction.

For rough contact $0 \leqslant F \leqslant \mu R$ and for smooth contact $F = 0$.

When friction is limiting, the resultant of μR and R is at an angle λ to R where λ is the angle of friction and $\tan \lambda = \mu$.

Work

When a constant force acts on an object the amount of work done by the force is given by

component of force in the direction of motion
× distance moved by object

The unit of work is the joule (J), which is the amount of work done when a force of 1 newton moves an object through 1 m.

The work done by a vehicle moving at constant speed is

driving force × distance moved

Power

Power is the rate at which work is done and is measured in watts where

1 watt (W) is 1 joule per second.

The power of a moving vehicle is the rate at which the driving force is working and this is given by

driving force × velocity

Energy

Energy is the ability to do work.

Energy and Work are interchangeable so energy is measured in joules.

The Kinetic Energy (KE) of a moving object is given by $\frac{1}{2}mv^2$; it can never be negative.

Potential Energy (PE) is equivalent to work done by gravity and is given by mgh where h is the height of an object above a chosen level.
PE is negative for a body below the chosen datum.

If the total work done by the external forces acting on a body is zero the total mechanical energy of the body remains constant.

Work done by gravity is accounted for as potential energy so weight is not included as an external force.

MISCELLANEOUS EXERCISE B

In this exercise use $g = 9.8$ unless another instruction is given

1. The diagram shows three coplanar forces of magnitudes 2 N, 3 N and P N all acting at a point O in the directions shown.

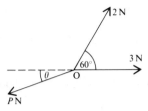

Given that the forces are in equilibrium obtain the numerical values of $P \cos \theta$ and $P \sin \theta$ and hence, or otherwise, find $\tan \theta$ and P. (AEB)

2. A particle of mass 2 kg is suspended from a point C of a light string, the ends of the string being attached to fixed points A and B. When the particle hangs in equilibrium, AC is horizontal and CB is inclined at 20° to the horizontal.

Find the magnitude of the tension in the string in

(a) the section BC   (b) the section AC   (AEB)

3. A particle of mass 2 kg which is free to move along the positive x-axis is at rest at time $t = 0$ s. For the first ten seconds it is acted upon by a force in the positive x direction of magnitude 6 N; for the next 240 seconds no force acts; then the particle is brought to rest by applying a force of magnitude 10 N along the x-axis. Find the time at which the particle comes to rest and the distance travelled up to that time. (AEB)

4. Two particles A and B are placed on a rough horizontal table at a distance a apart. The coefficient of friction between the table and A is $\frac{1}{2}$, and the coefficient of friction between the table and B is $\frac{3}{4}$.
The particles are projected simultaneously with velocity u in the direction AB. Given that the particles do not collide, find, in terms of u and g, the distance travelled by each particle before it comes to rest.

Deduce that $u^2 < 3ga$. (NEAB)

5. In a race, an athlete of mass 70 kg starts from rest and runs a distance of 100 m along a straight horizontal track. During the first 6 seconds of the race, the net propelling force is horizontal and of magnitude 175 N.
For the remainder of the race, the net force is a horizontal resistance of magnitude of F N.

(a) Find the speed of the athlete after the first 6 seconds.

(b) Given that the athlete completes the race in 11 s, find the value of F.
                                                           (AEB)

6. The points A, B and C of a horizontal plane have coordinates (4, 3), (−4, 0) and (4, −3), respectively, these dimensions being in metres. A particle P on the plane is subject to three forces which are directed towards A, B and C.

(a) When P is at the origin the forces directed towards A, B and C have magnitudes 4 N, 2 N and 4 N, respectively, as shown in the diagram.

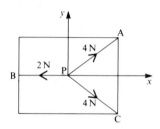

Calculate the magnitude of the resultant of the three forces, and state its direction.
   (i) Given that the plane is rough, and that P is in equilibrium at the origin, state the magnitude and direction of the frictional force on P.
   (ii) Given that the plane is smooth and that the mass of P is 0.1 kg, calculate the acceleration of P when it is at the origin.

(b) When P is at the point (−4, 3) the force directed towards B is zero, and the forces directed towards A and C have magnitudes 10 N and 14 N, respectively. Calculate the magnitude and direction of the resultant of the two non-zero forces.   (UCLES)

7.

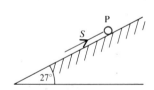

A small parcel P, of mass 1.5 kg, is placed on a rough plane inclined at an angle of 27° to the horizontal. The coefficient of friction between the parcel and the plane is 0.3. A force S, of variable magnitude, is applied to the parcel as shown in the diagram. The line of action of S is parallel to a line of greatest slope of the inclined plane.
Determine, in N to 1 decimal place, the magnitude of S when the parcel P is in limiting equilibrium and on the point of moving

(a) down the plane   (b) up the plane.   (ULEAC)

8. A light inextensible string passes over a small fixed smooth pulley. The string carries a particle of mass 0.06 kg at one end and a particle of mass 0.08 kg at the other end. The particles move in a vertical plane, with both hanging parts of the string vertical. Find the magnitude of the acceleration of the particles and the tension in the string.   (AEB)

9. A smooth pulley is fixed at a height $3l$ above a horizontal table and a light inextensible string hangs over the pulley. A particle of mass m is attached to one end of the string and a particle of mass $2m$ is attached to the other end. The system is held at rest with the particles hanging at the same level and at a distance l from the table. The parts of the string not in contact with the pulley are vertical. The system is then released from rest. Find, in terms of g and l, the speed u with which the particle of mass $2m$ strikes the table.

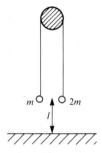

10. The diagram shows a box B, of mass 0.4 kg, resting on the smooth horizontal surface of a table. An inextensible light string connects B to another particle A of mass 0.1 kg. The string passes over a small smooth pulley fixed at the edge of the table. The system is released from rest with the string taut and the hanging part vertical.

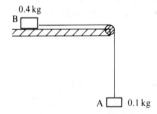

(a) Calculate the magnitude, in m/s², of the acceleration of B.

(b) Calculate the tension, in N, in the string.

(c) What would happen if after some time the string broke?   (ULEAC)

11. A long light inextensible string passes over a smooth pulley, and particles of masses m and $4m$ are fixed to the two ends of the string. The system is released from rest with the string taut and with each particle at a height 1.2 m from the floor. (Take g to be 10).

(a) Show that the acceleration of either particle is 6 m/s².

(b) Calculate the time taken for the heavier particle to reach the floor, and the speed on impact.

(c) Assuming the heavier particle does not rebound, calculate the greatest height above the floor attained by the lighter particle in the subsequent motion.

                                                                    (UCLES)

12.

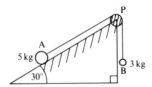

The diagram shows a particle A, of mass 5 kg, resting on a smooth plane which is inclined at 30° to the horizontal. A light inextensible string connects A to a second particle B, of mass 3 kg, which hangs freely. The string passes over a small smooth pulley P fixed at the top of the inclined plane, and the portion AP of the string is parallel to a line of greatest slope of the plane. The system is released from rest with the string taut and the hanging part vertical.

(a) Calculate, in m/s² to 2 decimal places, the acceleration of A.

(b) Calculate, in N to 2 decimal places, the tension in the string.   (AEB)

13.

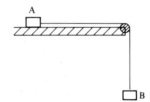

The diagram shows two particles, A of mass $3m$ and B of mass $2m$, connected by a light inextensible string which passes over a smooth fixed pulley at the edge of a horizontal table. Initially A is held at rest on the table and B is hanging freely at a height h above the floor. The particle A is then released and during its motion along the table experiences a retarding force, due to friction with the table, of magnitude $\frac{1}{3}\,mg$. The particle B strikes the floor before A reaches the edge of the table. Find, in terms of m and g, the tension in the string and the acceleration of the particles whilst they are both moving. Show that the speed of A at the instant when B hits the floor is $\sqrt{\left(\dfrac{2gh}{3}\right)}$.

When B hits the floor, A continues moving along the table but eventually comes to rest before it reaches the edge. Show that the length of the string must be greater than $4h$.   (NEAB)

Each question from 14 to 17 is followed by several suggested responses. Choose which is the correct response.

14. A particle travelling in a horizontal straight line has an acceleration of $+2$ m/s².

   A   Its total mechanical energy is constant.

   B   Its kinetic energy is constant.

   C   Work is being done on the particle.

15. A particle is moving with uniform velocity.

    A   The particle is in equilibrium.

    B   The particle has a constant acceleration.

    C   There is a resultant force acting on the body in the direction of the velocity.

16. Forces represented by $2\mathbf{i} + 5\mathbf{j}$, $\mathbf{i} - 8\mathbf{j}$, and $p\mathbf{i} + q\mathbf{j}$, are in equilibrium, therefore

    A $p = 3$ and $q = -3$ **C** $p = -2$ and $q = 3$

    B $p = -3$ and $q = 3$ **D** $p = 2$ and $q = -40$

17. The potential energy of a body of mass m is mgh where h is

    A   the distance from a chosen point

    B   the height above the ground

    C   the height above a chosen level.

18. A lorry has mass 6 tonnes (6000 kg) and its engine can develop a maximum power of 10 kW. When the speed of the lorry is v m/s the total non-gravitational resistance to motion has magnitude $25v$ N. Find the maximum speed of the lorry when travelling along a straight horizontal road. Find also the maximum speed up a hill which is inclined at an angle α to the horizontal where $\sin \alpha = \frac{1}{100}$, giving your answer to 2 decimal places.   (AEB)

19.

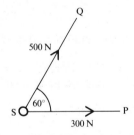

A heavy stone S, resting on rough horizontal ground, is to be moved by pulling horizontally on two ropes, SP and SQ, which are attached to the stone. The tensions in the ropes are 300 N and 500 N respectively, and the angle between the ropes is 60°, as shown in the diagram. Find, either by accurate drawing or by calculation, the resultant force on the stone due to the two ropes, giving its magnitude and the direction that it makes with SP.

The stone is dragged slowly along the ground for a distance of 2 m in the direction of the resultant pull, all forces remaining constant in magnitude and direction. Find the total amount of work done by the forces dragging the stone along the ground.

Assuming that the coefficient of friction between the stone and the ground is 0.5, find the mass of the stone. (Take g to be 10 m/s^2.)   (UCLES)

20. A lorry, travelling at constant speed, moves a distance of 2 km up a hill which is inclined at 10° to the horizontal. Given that the mass of the lorry is 2400 kg, and that the frictional resistance to the motion is of magnitude 800 N, find, in J, to 3 significant figures, the total work done by the engine of the lorry against the resistance and gravity.   (AEB)

21. A smooth bead P, of mass 0.15 kg, is threaded on a smooth straight fixed wire which is inclined at 35° to the horizontal. The bead P is released from rest and moves down the wire, starting from a point A. After 2 seconds, P passes through a point B on the wire. Calculate

(a) the distance AB, giving your answer in metres to one decimal place,

(b) the kinetic energy of P at B, giving your answer in joules to 1 decimal place.   (AEB)

22.

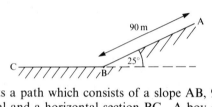

The diagram represents a path which consists of a slope AB, 90 m long, inclined at 25° to the horizontal and a horizontal section BC. A boy on a skate-board starts from rest at A and glides down AB before coming to rest between B and C. The magnitude of the resistive forces opposing the motion are constant throughout the journey. The combined mass of the boy and skate-board is 40 kg and the boy reaches B with a speed of 14 m/s. Calculate, to 3 significant figures,

(a) the energy lost, in J, by the boy and skate-board in going from A to B

(b) the magnitude, in N, of the resistive forces

(c) the distance, in m, the boy travels along BC before coming to rest.   (ULEAC)

23. A cyclist working at a rate of 150 watts maintains a steady speed of 27 km/h along a straight horizontal road. Find the resistance to motion, stating your units.   (AEB)

The instruction for answering questions 24 to 31 is: if the following statement must *always* be true, write T, otherwise write F.

24. If a frictional force acts on a body, it is not necessarily of value μR where R is the normal contact force.

25. The angle of friction is the angle between the frictional force and the normal reaction.

26. If a body has a resultant force acting on it the body will accelerate in the direction of the force.

27. A particle is hanging freely attached to a light inextensible string. The string is made to accelerate vertically upward. The tension in the string is greater than the weight of the particle.

28. A car is towing a van and is accelerating. The tension in the tow rope is greater than the resistance to the motion of the van.

29. If a particle has a constant acceleration it must be moving in a straight line.

30. As long as no external forces act on a system the kinetic energy must be constant.

31. Some external forces which act on a moving body do not do any work.

32. A lorry of mass 5000 kg is travelling up a slope inclined at an angle α to the horizontal, where $\sin \alpha = \frac{1}{200}$. The engine of the lorry is working at a steady 20 kW and the constant resistances due to friction and to airflow around the lorry amount to 600 N. At the instant when the lorry is moving at a speed of 12 m/s, calculate, in m/s^2 to 3 significant figures, the acceleration of the lorry.
                                                                                      (ULEAC)

33. The total mass of a train is 400 tonnes. It moves on a straight horizontal track against a constant resistance of 20 kN. Find the rate, in kilowatts, at which the engine is working when it is travelling at a uniform speed of 63 km/h.
   The tractive force of the engine is now increased to 25 kN and maintained at this value. The resistance remains unchanged. As a result, the speed of the train increases uniformly from 63 km/h to 81 km/h. For this part of the motion, show that the acceleration of the train is $\frac{1}{80}$ m/s^2 and find

   (a) the time taken

   (b) the distance travelled.   (NEAB)

34. The resistance to the motion of a motor coach is K newtons per tonne, where K is a constant. The motor coach has mass $4\frac{1}{2}$ tonnes. When travelling on a straight horizontal road with the engine working at 39.6 kW, the coach maintains a steady speed of 40 m/s.

   (a) Show that $K = 220$.

   The motor coach ascends a straight road, which is inclined at an angle α to the horizontal, where $\sin \alpha = 0.3$, with the same power output and against the same constant resisting forces.

   (b) Find, in joules to 2 significant figures, the kinetic energy of the motor coach when it is travelling at its maximum speed up the slope.   (ULEAC)

35. The resistance to motion of a car of mass 2000 kg is proportional to its speed. With the engine working at 72 kW the car can attain a maximum speed of 12 m/s when travelling up a straight road, which is inclined at an angle α to the horizontal where $\sin \alpha = \frac{5}{49}$.

    (a) Show that the resistance to motion is 4000 N at this speed.

    (b) Find the greatest speed at which the car could travel *down* this road with the engine working at 72 kW.   (AEB)

36. A motor cyclist together with his machine has a mass of 200 kg. He is ascending a straight road inclined at θ to the horizontal where $\sin \theta = \frac{1}{7}$ against a constant resistance of 120 N and the engine is working at 9 kW. Calculate

    (a) the acceleration of the motor cyclist when his speed is 20 m/s

    (b) the maximum speed which he can attain up the incline.

    Later the motor cyclist descends the same hill with a pillion passenger whose mass is 75 kg. Given that the constant resistance is now 165 N and that the engine is switched off, find the time taken and the distance covered as the speed of the motor cyclist increases from 10 m/s to 20 m/s.   (WJEC)

37. A car, of mass M kilograms, is pulling a trailer, of mass λM kilograms, along a straight horizontal road. The tow-bar connecting the car and the trailer is horizontal and of negligible mass., The resistive forces acting on the car and trailer are constant and of magnitude 300 N and 200 N respectively. At the instant when the car has an acceleration of magnitude 0.3 m/s^2, the tractive force has magnitude 2000 N.

    Show that $M(\lambda + 1) = 5000$.

    Given that the tension in the tow-bar is 500 N at this same instant, find the value of M and the value of λ.   (ULEAC)

38. The total mass of a woman and her bicycle is 80 kg. The woman freewheels down a slope inclined at an angle θ to the horizontal, where $\sin \theta = 0.14$, with constant acceleration of magnitude 0.9 m/s^2.

    (a) Prove that the total magnitude of the resistive forces opposing the motion is 37.76 N.

    (b) Find the time required for the woman to cover 180 m from rest.

    The woman cycles up the same slope at constant speed 6 m/s the resistive forces remaining unchanged.

    (c) Find, in W to 3 significant figures, the power that must be exerted by the woman.

    (d) If now the woman suddenly increases her work rate by 240 W, find the magnitude of her acceleration up the slope, to 3 significant figures, at that instant.   (ULEAC)

39. (In this question take g to be 10)

(a) A weightlifter lifts a weight of mass 100 kg from the floor to a height of 2 m above the floor. Calculate the work done on the weight by the weightlifter.
The weightlifter then allows the weight to fall back to the floor. State the loss in potential energy of the weight, and hence calculate the speed of the weight on impact with the floor.

(b) Water flows over a waterfall where there is a vertical drop of 80 m. The water at the top of the waterfall is flowing at a speed of 3 m/s. By considering the potential and kinetic energy of 1 kg of water, or otherwise, find the speed of the water after it has fallen 80 m.

Water flows over the waterfall at a rate of 200 m³/s and 1 m³ of water has a mass of 1000 kg. Assuming that 40% of the energy of the water at the bottom of the waterfall can be converted into electricity by suitable generators, calculate the power, in kilowatts, that could be developed.

(UCLES)

CHAPTER 9

PROJECTILES

THE MOTION OF A PROJECTILE

A particle that is sent moving into the air is called a projectile. At this stage we consider that air resistance is small enough to be ignored so the only force acting on a projectile, once it has been thrown, is its own weight. It follows that the projectile has an acceleration downward of magnitude g, but has constant velocity horizontally.

Therefore the horizontal motion and the vertical motion are analysed separately.

The horizontal and vertical components of velocity and displacement will be involved and a new form of notation is used to denote these quantities.

Consider a particle projected with a velocity V at an angle θ to the horizontal. We take O as the point of projection, Ox as a horizontal axis and Oy as a vertical axis.

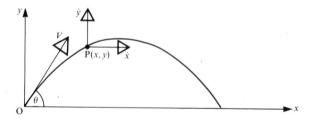

At any time during its flight, the projectile is at a point where:

the horizontal displacement from O is x

and the vertical displacement from O is y

Now the horizontal velocity is the rate at which x increases and this is denoted by $\dot{x}$. (The dot over the x means 'the rate of increase with respect to time'.)

Similarly the vertical velocity at any time is denoted by $\dot{y}$.

The initial components of velocity are:

$$V \cos \theta \quad \text{in the direction } Ox$$
$$V \sin \theta \quad \text{in the direction } Oy.$$

Considering the horizontal motion where the velocity is constant:

At any time t seconds after projection,

$$\dot{x} = V \cos \theta \qquad \qquad [1]$$
$$x = (V \cos \theta) \times t = Vt \cos \theta \qquad \qquad [2]$$

Considering the vertical motion where there is an acceleration g downwards:
Using $v = u + at$ and $s = ut + \frac{1}{2}at^2$ where $u = V \sin \theta$, gives

$$\dot{y} = V \sin \theta - gt \qquad \qquad [3]$$
$$y = (V \sin \theta) \times t - \frac{1}{2}gt^2 = Vt \sin \theta - \frac{1}{2}gt^2 \qquad \qquad [4]$$

These four equations provide all the information needed to investigate the motion of any projectile.

If, for example, we want the speed v of the particle at a particular time, $\dot{x}$ and $\dot{y}$ can be found from equations [1] and [3], then $v^2 = \dot{x}^2 + \dot{y}^2$.

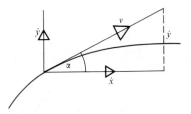

Also, the direction of motion, i.e. the way the particle is moving, is given by the direction of the velocity.

So if the direction of motion makes an angle α with the horizontal, then

$$\tan \alpha = \dot{y}/\dot{x}$$

Examples 9a

1. A particle P is projected from a point O with a speed of 40 m s^{-1}, at 60° to the horizontal. Find, 3 seconds after projection,

(a) the speed of the particle

(b) the horizontal and vertical displacements of the particle from O

(c) the distance of P from O.

Take g as 10 and give answers corrected to 2 significant figures.
Note that m s^{-1} is an alternative form for m/s.

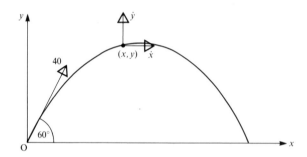

(a) Use equations [1] and [3] with $V = 40$, $\theta = 60°$ and $t = 3$.

Horizontally, $\dot{x} = 40 \cos 60° = 20$

Vertically, $\dot{y} = 40 \sin 60° - gt = 34.64 - 30$

$\Rightarrow$ $\dot{y} = 4.64$

If v m s^{-1} is the speed of the particle after 3 seconds,

$v^2 = 4.64^2 + 20^2$ $\Rightarrow$ $v = 20.53\ldots$

The speed of the particle is 21 m s^{-1} (2 sf)

(b) Use equations [2] and [4]

Horizontally, $x = 40 \times 3 \cos 60° = 60$

Vertically, $y = 40 \times 3 \sin 60° - \frac{1}{2}gt^2 = 103.92 - 45$

$y = 58.92$

The displacements of the particle from O are:

60 m horizontally and 59 m vertically (2 sf)

(c) $OP^2 = x^2 + y^2$

$= 60^2 + 58.92^2 = 7072$

The distance of P from O is 84 m (2 sf)

2. The initial velocity of a particle projected from a point O is $u\mathbf{i} + v\mathbf{j}$ where $\mathbf{i}$ and $\mathbf{j}$ are horizontal and vertical unit vectors respectively.
Find, in terms of u, v, g and t

(a) the velocity vector of the particle t seconds after projection

(b) the position vector of the particle at this time.

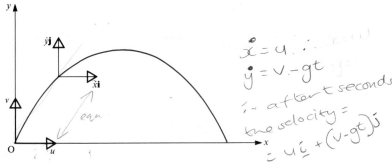

$\dot{x} = u$

$\dot{y} = v - gt$

after t seconds
the velocity =
$= u\mathbf{i} + (v - gt)\mathbf{j}$

In the direction Ox, $\dot{x}$ is constant and equal to u

and $x = ut$

In the direction Oy, $\dot{y} = v - gt$

and $y = vt - \frac{1}{2}gt^2$

$x = ut$

$y = vt - \frac{1}{2}gt^2$

disp $=$
$ut\mathbf{i} + (vt - \frac{1}{2}gt^2)\mathbf{j}$

$\therefore$ after t seconds

(a) the velocity of the particle is $u\mathbf{i} + (v - gt)\mathbf{j}$

(b) the position of the particle is $ut\mathbf{i} + (vt - \frac{1}{2}gt^2)\mathbf{j}$

3. A particle P is projected from a point 5 m above the ground. The horizontal and vertical components of the velocity of projection are each 24 m s^{-1}.

(a) Find the angle of projection.

(b) Taking g as 10 find the horizontal distance of P from the point of projection when it hits the ground.

$x = 24 \cos 45 \times$

$\sqrt{u^2 + v^2}$

$u \times u + v \times v$

$u + v$

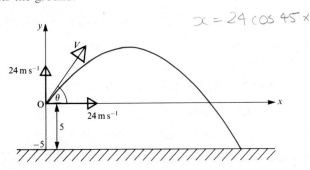

(a) If the velocity of projection is V then

$$V \cos \theta = 24 \quad \text{and} \quad V \sin \theta = 24$$

$$\therefore \qquad \tan \theta = \frac{V \sin \theta}{V \cos \theta} = \frac{24}{24} = 1$$

$$\Rightarrow \qquad \theta = 45°$$

(b) When the particle hits the ground it is 5 m *below* the point of projection so $y = -5$.

At any time t, $\qquad y = 24t - \tfrac{1}{2}gt^2$

$\therefore \qquad$ when $\qquad y = -5, \quad -5 = 24t - 5t^2 \qquad \Rightarrow \qquad 5t^2 - 24t - 5 = 0$

Hence $\qquad\qquad (5t + 1)(t - 5) = 0 \qquad \Rightarrow \qquad t = -\tfrac{1}{5}$ or 5

A negative time has no meaning in this problem, so $t = 5$.

At any time t, $x = 24t$ so, when $t = 5, \ x = 120$

When P hits the ground its horizontal distance from O is 120 m.

4. The top of a tower is 10 m above horizontal ground. A boy fires a stone from a catapult with a velocity of 12 m s^{-1}. Find how far from the foot of the tower the stone hits the ground if

(a) it is fired horizontally

(b) it is fired at 30° below the horizontal.

Take g as 10 and give answers corrected to 2 significant figures.

(a)

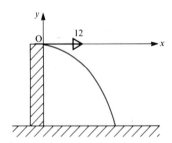

When the initial velocity is horizontal, $\theta = 0$, therefore $\sin \theta = 0$ and $\cos \theta = 1$.

At any time t, $\qquad x = Vt \quad$ and $\quad y = -\tfrac{1}{2}gt^2$

When the stone hits the ground, $\quad y = -10$

$\therefore \qquad\qquad -\tfrac{1}{2}gt^2 = -10 \qquad \Rightarrow \qquad t^2 = \tfrac{10}{5} = 2$

$\therefore \qquad\qquad t = \sqrt{2} \ (\ t$ cannot be negative$)$

When $\quad t = \sqrt{2}, \qquad x = Vt = 12\sqrt{2} = 16.97\ldots$

The stone hits the ground 17 m from the foot of the tower (2 sf).

(b)

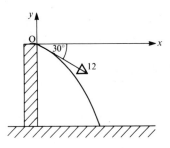

The stone is fired at a downward angle so θ is a negative acute angle.

At any time t, $\quad x = Vt \cos(-30°) = Vt \cos 30°$

and $\qquad\qquad y = Vt \sin(-30°) - \frac{1}{2}gt^2 = -Vt \sin 30° - \frac{1}{2}gt^2$

When the stone hits the ground, $y = -10$

$\therefore \qquad 12t(-\frac{1}{2}) - \frac{1}{2}gt^2 = -10 \qquad \Rightarrow \qquad 5t^2 + 6t - 10 = 0$

$$\Rightarrow \qquad t = \frac{-6 \pm \sqrt{236}}{10}$$

$\therefore \qquad t = 0.936 \quad (t \text{ cannot be negative})$

When $\quad t = 0.936, \quad x = Vt \cos 30° = 12 \times 0.936 \times 0.8660$

$$= 9.726\ldots$$

The stone hits the ground 9.7 m from the foot of the tower (2 sf).

Note that it is possible to solve problems in which θ is negative, without referring to the standard projectile equations, but by starting from first principles taking the downward direction as positive. The reader is given an opportunity to try this approach in question 25 in the next exercise.

5. A particle P is projected from a point O with velocity V at an angle θ to the horizontal. Find, in terms of V and θ, the time at which P reaches its greatest height, and what the greatest height is.

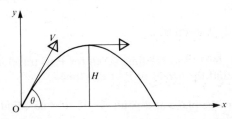

When P is at its greatest height it is momentarily travelling horizontally, i.e. $\dot{y} = 0$.

$$\dot{y} = V \sin \theta - gt$$

$\therefore$ when $\dot{y} = 0 \qquad V \sin \theta = gt \qquad \Rightarrow \qquad t = \dfrac{V \sin \theta}{g}$

i.e. the greatest height is reached after a time of $\dfrac{V \sin \theta}{g}$

The height at time t is given by $\qquad y = Vt \sin \theta - \tfrac{1}{2}gt^2$

$\therefore$ for the greatest height $H \qquad H = V\left(\dfrac{V \sin \theta}{g}\right) \sin \theta - \tfrac{1}{2}g\left(\dfrac{V \sin \theta}{g}\right)^2$

$$= \dfrac{V^2 \sin^2 \theta}{g} \left(1 - \tfrac{1}{2}\right)$$

$\therefore$ the greatest height reached is $\qquad \dfrac{V^2 \sin^2 \theta}{2g}$

EXERCISE 9a

Unless another instruction is given, take g as 10 m s^{-2} (i.e. m/s^2) and give answers corrected to 2 significant figures; these approximations are justified as we have already made one approximation in ignoring air resistance.
In questions where vector notation is used, **i** and **j** are horizontal and vertical unit vectors respectively.

In each question from 1 to 5, a particle P is projected from a point O on a horizontal plane with velocity V at an angle θ to the horizontal. All units are based on metres and seconds.

1. Given that $V = 24.5$ and $\theta = 30°$, find the speed of P after

   (a) 1 second   (b) 2 seconds.

2. If $V = 20$ and $\theta = 60°$, find the height of P above the plane when

   (a) $t = 1$ (b) $t = 2$ (c) $t = 3$.

   Illustrate your answers with a sketch.

3. Given that $V = 30$ and $\tan \theta = \frac{3}{4}$, find the speed and the coordinates of the position of P after

   (a) 1 second   (b) 2 seconds.

4. If $V = 10$ and $\theta = 60°$, find the time taken for P to travel a horizontal distance of 5 m and find the height of P at this time.

5. After 4 seconds P hits the plane at a point A. If $\theta = 45°$, find

   (a) V (b) the distance OA.

6. A particle is projected with a velocity vector $20\mathbf{i} + 40\mathbf{j}$. Find the velocity vector of the particle after 3 seconds.

7. A particle P is projected from a point O with a velocity of 10 m/s at an angle of 30° to the horizontal. Find the horizontal and vertical displacements of P from O after half a second and hence find the distance from O to P at this time.

8. A stone is thrown from a point O on the top of a cliff of height 60 m, and falls into the sea $2\frac{1}{2}$ seconds later at a point whose displacement from O is $24\mathbf{i} - 60\mathbf{j}$. Find, in the form $a\mathbf{i} + b\mathbf{j}$, the velocity of projection.

9. A particle is projected from a point O with a velocity vector $\mathbf{i} + 2\mathbf{j}$. Find the velocity and the position of the particle in vector form after

   (a) t seconds   (b) $1\frac{1}{2}$ seconds.

10. A particle P is projected from a point O with velocity **V**. Giving answers in terms of g find, in the form $a\mathbf{i} + b\mathbf{j}$, the velocity and position of P after t seconds if

    (a) $\mathbf{V} = 5\mathbf{i} + 3\mathbf{j}$

    (b) $\mathbf{V} = 4\mathbf{i} - \mathbf{j}$

    (c) $\mathbf{V} = 20\,\mathrm{m\,s^{-1}}$ at 60° to the horizontal

    (d) $\mathbf{V} = 10\sqrt{5}\,\mathrm{m\,s^{-1}}$ at θ to the horizontal where $\tan\theta = 2$.

11. At fielding practice a cricketer throws the ball with a speed of 26 m/s at an angle α above the horizontal, where $\tan\alpha = \frac{12}{5}$. Find

    (a) the times at which the ball is 16 m above the ground

    (b) the horizontal distance covered between these times.

12. A stone is thrown downwards from a point A into a quarry that is 25 m deep. Find the initial speed and the direction of projection if, after 2 seconds, the stone lands at the bottom of the quarry at a horizontal distance from A of

    (a) 30 m   (b) 20 m.

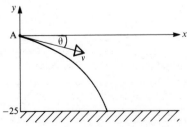

13. A particle is projected with a velocity of $20\sqrt{2}$ m s^{-1} at an angle of $45°$ to the horizontal. Find, in terms of g, how long it is before the particle reaches its highest point and what the greatest height is.

14. A particle is projected from a point O and after $1\frac{1}{2}$ seconds it passes through a point whose position is represented by the vector $4\mathbf{i} + \mathbf{j}$. Find the initial velocity as a vector.

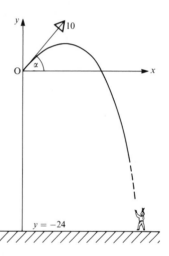

15. A boy is at the window of his flat and throws a ball to his friend on the ground 24 m below. The ball is thrown with speed 10 m/s at an angle of elevation α, where $\tan \alpha = \frac{5}{12}$.

   (a)   Find how long it takes for the ball to reach ground level.

   (b)   If the friend is standing 8 m from the wall of the flat, will he be able to catch the ball without moving?

16. A particle is projected with a velocity vector $a\mathbf{i} + b\mathbf{j}$. After 5 seconds the velocity vector is $40\mathbf{i} + 60\mathbf{j}$. Find the values of a and b.

17. A particle P is projected from a point O with a velocity of 25 m s^{-1} at an angle of $60°$ to the horizontal. For how long is P at least 15 m above the level of O?

18. A golf ball is hit by a golfer at 25 m/s towards the green which is 2 m below the level of the tee. The ball is struck at an angle θ to the horizontal where $\tan \theta = \frac{7}{24}$ and lands directly in the hole! What is the horizontal distance from tee to hole?

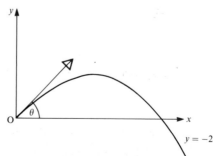

19. In a Highland Games Competition the local strongman, Mac, is hoping to break the record for throwing the heavy hammer. He manages to hurl the hammer with a speed of 25 m s^{-1} at $40°$ to the horizontal. If the record throw is 60.6 m, does Mac break the record?

20. A stone is thrown from the top of a cliff at an angle α to the horizontal and with a speed of 19.5 m/s. The stone falls into the water 37.5 m from the foot of the cliff. Given that $\tan \alpha = \frac{12}{5}$, find the height of the cliff if α is

(a) above the horizontal

(b) below the horizontal.

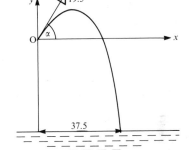

21. Find the angle of projection of a ball, thrown at 20 m s^{-1}, which is at its greatest height when it *just* passes over the top of a tree that is 16 m high.

22. From the top of a fifty-metre high cliff, the angle of depression of a marker buoy is $30°$. A student throws a stone from the cliff top in an attempt to hit the buoy but is foolish enough to think that the stone should be thrown at $30°$ below the horizontal. How far short of the buoy will the stone land if it is thrown with a speed of 15 ms^{-1}? 54 m as answer

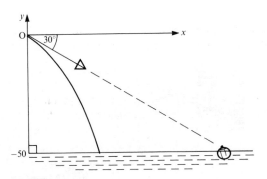

23. The horizontal and vertical components of the initial velocity of a projectile are each of magnitude u. Express in the form $a\mathbf{i} + b\mathbf{j}$

(a) the initial velocity

(b) the velocity of the projectile after (i) t seconds (ii) 2 seconds

(c) the position of the projectile after 2 seconds.

24. An arrow is shot from the top of a building 26 m high. The initial speed of the arrow is 30 m s^{-1}. Find how long the arrow is in the air if it is fired at an angle of 20° to the horizontal

(a) upwards   (b) downwards.

25. A particle P is projected from a point O, with speed V at an angle θ *below* the horizontal.

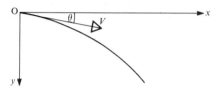

Taking the axes shown in the diagram find, t seconds after projection, an expression in terms of g for

(a) the horizontal velocity component

(b) the vertical velocity component

(c) the horizontal displacement of P from O

(d) the vertical displacement of P from O.

26. A particle P is projected from a point O, with speed 20 m s^{-1} at an angle of 30° below the horizontal. Use the results found in question 25 to find

(a) the speed of P

(b) the distance of P from O, after 2 seconds.

27.

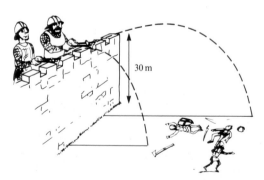

From the battlements of a castle, 30 m above the ground, defenders are catapulting rocks at their attackers below. They all think that the best angle of projection is 20°. The strongest catapulter can fire a rock with a speed of 6 m/s and even the weakest fires at 2 m/s. Find the distances from the foot of the castle walls between which the attackers may be hit.

PROPERTIES OF THE FLIGHT PATH OF A PROJECTILE

The Equation of the Path

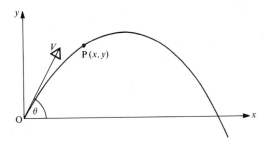

Using x and y axes through the point of projection, we have seen that the coordinates of the position of the projectile at any time can be expressed as

$$x = Vt \cos \theta$$
$$y = Vt \sin \theta - \tfrac{1}{2}gt^2$$

As these two equations give x and y each in terms of the variable t, they are the parametric equations of the path of the projectile (which is also called the trajectory).

From the first equation, $t = \dfrac{x}{V \cos \theta}$

Substituting in the second equation gives

$$y = \frac{Vx \sin \theta}{V \cos \theta} - \tfrac{1}{2}g\left(\frac{x}{V \cos \theta}\right)^2$$

$\Rightarrow$ $\qquad\qquad\qquad y = x \tan \theta - \dfrac{gx^2}{2V^2 \cos^2 \theta}$ $\qquad\qquad$ [1]

For any particular projectile, g, V and θ are constants, so y is a quadratic function of x, showing that the path of a projectile is a parabola with a vertical axis of symmetry.

Equation [1] is called *the equation of the path* of the projectile, or the equation of the *trajectory*.
It is very useful in problems where the position of the projectile is involved but the time taken to reach the position is not.

In particular if the angle of projection, θ, has to be found, the equation of the path can be rearranged as follows:

Using $\dfrac{1}{\cos^2 \theta} = \sec^2 \theta = \tan^2 \theta + 1$ gives the equation of the path as

$$y = x \tan \theta - \frac{gx^2}{2V^2} (\tan^2 \theta + 1)$$

If we are given the coordinates of a point through which the projectile passes, and V is known, this becomes a quadratic equation in $\tan \theta$.
The reader should be prepared to use the equation of the trajectory in either of the forms above to solve a problem in which time does not appear.

Note that the equation is valid for the whole of the path, including those cases where the projectile moves on below the level of the point of projection, e.g. a stone thrown from the top of a cliff.

The Greatest Height Reached

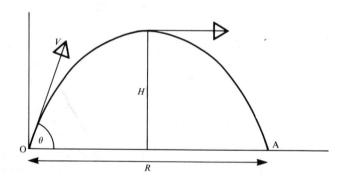

In Examples 9a, we saw in example 5 that, for a projectile given an initial velocity V at an angle θ to the horizontal, the greatest height H reached above the point of projection is given by

$$H = \frac{V^2 \sin^2 \theta}{2g}$$

Although this expression can sometimes be *quoted* in a solution, its derivation may be asked for. So it is important to remember that it is found by using the time taken to reach the point where $\dot{y}$ is zero.

The Range on a Horizontal Plane

The range of a projectile on a horizontal plane is the distance between the point of projection O and the point where the projectile returns to the level of O. The range is usually denoted by R as shown in the diagram opposite.

When the projectile reaches A, $y = 0$,

i.e. $\quad Vt \sin \theta - \frac{1}{2}gt^2 = 0 \quad \Rightarrow \quad t = \dfrac{2V \sin \theta}{g}$

Note that this is the time taken for the whole of the journey; it is known as *the time of flight.*

Note also that the time of flight is twice the time taken to reach the highest point on the path.

Now R is the value of $x \ (= Vt \cos \theta)$ at this time,

i.e. $\quad R = V \cos \theta \left(\dfrac{2V \sin \theta}{g} \right) = \dfrac{V^2}{g} (2 \sin \theta \cos \theta)$

$$\therefore \qquad R = \dfrac{2V^2 \sin \theta \cos \theta}{g} \quad \text{or} \quad \dfrac{V^2 \sin 2\theta}{g}$$

The second of these forms is best if the problem is simply to find R, but when R is being used in conjunction with other distances, e.g. the greatest height, the first form is usually better.

Note that we are referring above to the range *on a horizontal plane,* i.e. the distance between O and the point where the projectile is again level with O.

If a projectile ends its flight above O (e.g. a ball that lands on a roof), or below O (e.g. a stone thrown from the top of a tower to the ground), the range is the distance between O and the landing point but is *not* given by the expressions above.

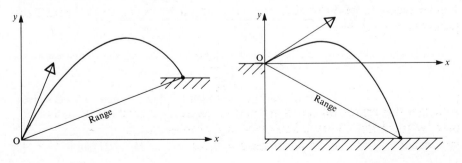

The Maximum Horizontal Range

The formulae derived above for the horizontal range of a projectile show that the value of R depends upon the initial velocity *and* the angle of projection.

Therefore, for a given value of V, the range varies only with θ and the maximum range occurs when $\sin 2\theta$ is maximum.

The greatest value of $\sin 2\theta$ is 1 and this is when $\quad 2\theta = 90° \quad \Rightarrow \quad \theta = 45°$

Therefore **the *maximum* horizontal range, R_{max}, is $\dfrac{V^2}{g}$,**

and **it is achieved when $\theta = 45°$**

Examples 9b

1. Two seconds after it is projected from a point O on a horizontal plane, a particle P passes through a point represented by $3\mathbf{i} + 5\mathbf{j}$, where $\mathbf{i}$ and $\mathbf{j}$ are respectively vectors of $1\ \mathrm{m\,s^{-1}}$ horizontally and vertically.

Find

(a) the speed and direction of projection

(b) the range of the projectile.

Take g as 10 and give answers corrected to 2 significant figures.

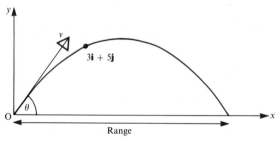

In this problem, although we have the coordinates of a point on the trajectory, we do not use the equation of the path, as time *is* involved.

(a) Using $\quad x = Vt \cos\theta \quad$ and $\quad y = Vt \sin\theta - \frac{1}{2}gt^2 \quad$ when $\quad t = 2 \quad$ gives

$$3 = 2V \cos\theta \qquad\qquad [1]$$

and $$5 = 2V \sin\theta - 20 \quad \Rightarrow \quad 25 = 2V \sin\theta \qquad\qquad [2]$$

[2] ÷ [1]　　　$\dfrac{2V \sin \theta}{2V \cos \theta} = \dfrac{25}{3}$　　　$\Rightarrow$　　　$\tan \theta = \dfrac{25}{3}$

$\Rightarrow$　　　　　　　　　$\theta = 83.15°, \quad \text{i.e.} \quad 83° \,(2 \text{ sf})$

From [1]　　$V = \dfrac{3}{2 \cos 83.15°} = 12.576\ldots$

The speed of projection is 13 m s^{-1} (2 sf)

(b)　　　　$\text{Range} = \dfrac{2V^2 \sin \theta \cos \theta}{g} = \dfrac{2(12.58)^2 \sin 83.15° \cos 83.15°}{10}$

$$= 3.748\ldots$$

The range is 3.7 m (2 sf).

2.　A gun is fired with a muzzle speed of 100 m s^{-1} in a tunnel whose roof is 4 m above the point of projection. Find the greatest permissible angle of projection if the bullet is to avoid hitting the roof. Find also the range of the gun with this angle of projection.

Take g as 10 and give answers corrected to 2 significant figures.

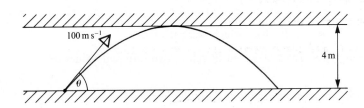

The greatest height of the bullet is 4 m above O.

As we are not asked to derive the expression for the greatest height, it can be quoted.

$H = \dfrac{V^2 \sin^2 \theta}{2g}$　　　$\Rightarrow$　　　$4 = 10^4 \sin^2 \theta \div 20$

$\therefore$　　　$\sin \theta = 0.08944$　　　$\Rightarrow$　　　$\theta = 5.13°, \quad \text{i.e.} \quad 5.1° \,(2 \text{ sf })$

The range R is given by　　$R = \dfrac{V^2 \sin 2\theta}{g}$

$\therefore$　　　　　　　$R = (10^4 \sin 10.26°) \div 10 = 180 \text{ m } (2 \text{ sf })$

3. A missile fired from a point O, with velocity 40 m s^{-1} at an angle α to the horizontal, passes through a point distant 32 m horizontally and 45 m vertically from O. Show that there are two possible angles of projection and give their values (take g as 10 and give your answers to the nearest degree). Illustrate your answers on a diagram.

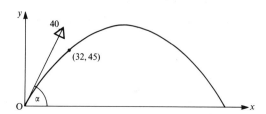

The coordinates of a point on the path of the missile are given and the time when the point is reached does not matter; so we use the equation of the path.

Using $\quad y = x \tan \alpha - \dfrac{gx^2 (\tan^2 \alpha + 1)}{2V^2}$

gives $\quad 45 = 32 \tan \alpha - \dfrac{10 \times (32)^2 \times (\tan^2 \alpha + 1)}{2 \times 1600}$

Using $\tan \alpha = T$, this equation simplifies to

$$45 = 32T - 3.2(T^2 + 1) = 0$$

$\Rightarrow \qquad\qquad 3.2T^2 - 32T + 48.2 = 0$

This is a quadratic equation for T in which $\quad a = 3.2, \quad b = -32, \quad c = 48.2$

$$b^2 - 4ac = 407.04, \quad \text{which is positive.}$$

So there are two values of T and therefore two different values of α.

Solving the equation by using the formula gives

$$T = \tan \alpha = \frac{32 \pm \sqrt{407.04}}{6.4} = \frac{32 \pm 20.18}{6.4} = 8.153 \quad \text{or} \quad 1.847$$

$\therefore$ the two values of α are $62°$ and $83°$.

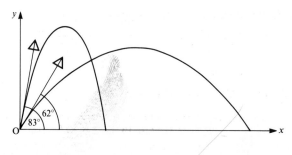

4. A particle is projected from a point O at ground level, at an angle of elevation of 55°. It just clears the top of each of two walls that are 2 m high. If the first of the walls is distant 2 m from O, find

(a) the speed of projection

(b) the distance of the second wall from O.

Take g as 10 and give answers corrected to 2 significant figures.

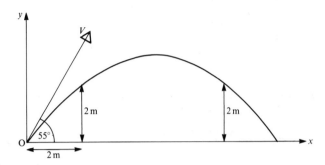

(a) As the projectile *just* clears the first wall we can take $(2, 2)$ as a point on the path.

Using $y = x \tan \theta - \dfrac{gx^2}{2V^2 \cos^2 \theta}$ gives

$$2 = 2 \tan 55° - \frac{10 \times 4}{2V^2 \cos^2 55°} = 2.856 - \frac{60.79}{V^2}$$

$\Rightarrow \qquad V^2 = 60.79 \div 0.856 = 71.02 \ldots \qquad \Rightarrow \qquad V = 8.427 \ldots$

The speed of projection is 8.4 m s^{-1} (2 sf).

(b) To find the location of the second wall we need the other value of x for which $y = 2$ so we use the equation of the path again.

When $y = 2$ and $V = 8.426$, the equation of the trajectory becomes

$$2 = 1.4281x - \frac{10x^2}{2 \times 71.02 \times 0.3290} = 1.428x - 0.2140x^2$$

$\Rightarrow \qquad\qquad x^2 - 6.673x + 9.346 = 0$ [1]

Solving this quadratic equation by formula gives

$$x = \frac{6.673 \pm \sqrt{(6.673^2 - 4 \times 9.346)}}{2} = \frac{6.673 \pm 2.673}{2}$$

$\therefore \qquad x = 2 \text{ or } 4.673$

The second wall is 4.7 m from O (2 sf).

Note that the second root of equation [1] could have been found directly, using the fact that one of the roots is 2 and also that the sum of the roots is $-b/a$. Hence $x + 2 = 6.673 \qquad \Rightarrow \qquad x = 4.673$.

5. A particle P is projected from a point O on a horizontal plane with speed V at an angle θ to the horizontal.

(a) State the value of θ for which the range is maximum and give the value of the maximum range.

(b) Find, in terms of V and g, the ratio of the greatest height to the maximum range.

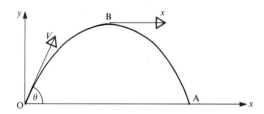

(a) The maximum range is achieved when $\theta = 45°$

$$R_{max} = \frac{V^2}{g}$$

(b) When the particle is at its greatest height, $\dot{y} = 0$

$$\therefore \qquad V \sin 45° - gt = 0 \qquad \Rightarrow \qquad t = \frac{V \sin 45°}{g}$$

The greatest height is the value of y when $t = (V \sin 45°)/g$

i.e. $\qquad y = Vt \sin 45° - \tfrac{1}{2}gt^2 = \dfrac{V^2 \sin^2 45°}{g} - \tfrac{1}{2}g\dfrac{(V^2 \sin^2 45°)}{g^2}$

$$\therefore \qquad H_{max} = \frac{V^2}{2g} - \frac{V^2}{4g} = \frac{V^2}{4g}$$

The ratio $\ H_{max} : R_{max}\ $ is $\ \dfrac{V^2}{4g} : \dfrac{V^2}{g} = 1 : 4$

EXERCISE 9b

For questions where quantities are given in the form $a\mathbf{i} + b\mathbf{j}$, $\mathbf{i}$ and $\mathbf{j}$ are respectively horizontal and vertical unit vectors, measured in the unit consistent with other quantities in the question.

In questions 1 to 6, a particle P is projected from a point O on a horizontal plane, with speed V and angle of elevation θ. The greatest height reached is H and the range on the plane is R. The maximum range is R_{max}. If you can *remember* the formulae for H and R, you may quote them but if not, derive them as quickly as you can. Do *not* rely on looking up each formula as you need it.

Throughout the exercise take g as 10 and give answers corrected to 2 significant figures unless another instruction is given.

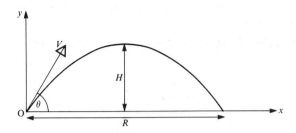

1. $V = 24$ m s^{-1} and $\theta = 30°$, find H and R.

2. $V = 20$ m s^{-1} and $R = 28$ m, find θ.

3. $\theta = 45°$ and $H = 10$ m, find V and R.

4. $R = 60$ m and $\theta = 60°$, find V and H.

5. $V = 30$ and R has its maximum value, find R_{max}.

6. $H = 16$ m and $V = 40$ m s^{-1}, find θ and R.

7. A stone is thrown with speed 26 m s^{-1} at an angle α to the horizontal where $\tan \alpha = \frac{5}{12}$. Find how far it has travelled horizontally when

(a) it is at the same level as the point of projection

(b) it is 2 m below the point of projection.

8. A particle is projected from a point O with speed 50 m s^{-1} at an angle of elevation of 40°. Taking Ox as the horizontal axis and Oy vertically upward, *find* (i.e. do not quote) the equation of the flight path of the particle. Hence find the height of the particle when it is distant 20 m horizontally from O.

9. A particle is projected from a point on a horizontal plane with a speed of 12 m/s. Find the angle of projection and the range if the time of flight is half a second.

10. Using a horizontal x axis and a vertical y axis find the equation of the path of a projectile whose initial velocity is

(a) 5 m s^{-1} at an angle θ above the horizontal where $\tan \theta = \frac{4}{3}$

(b) 30 m s^{-1} at an angle of 45° below the horizontal.

11. A ball is thrown with a speed of 16 m s^{-1} from a point on a level playground. The angle of projection is α where $\tan \alpha = \frac{3}{4}$. Find the time for which the ball is in the air and the horizontal distance it travels in this time.

12. A stone is thrown from ground level with a speed of 15 m/s, so that when it is travelling horizontally it just passes over a tree of height 3 m. Find the angle of projection.

13. A particle is projected from a point on a horizontal plane with speed $\sqrt{(2gh)}$. Find in terms of h, the range on the plane if the angle of projection is

(a) 30°   (b) 45°   (c) 60°.

14. A missile is fired at 80 m s^{-1} at an angle α to the horizontal. The missile must pass over an obstruction that is 20 m high and 120 m away in the line of flight. Find the smallest permissible value of α.

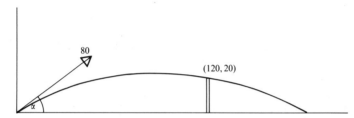

15. In a children's game played with large pebbles, the aim is to throw a pebble so that it lands within a circle of diameter 0.2 m, drawn on the ground. A player stands at A, a point marked on the ground, which is 3 m from Q as shown. One child can throw the stone with a speed of 5 m/s from a height of 0.6 m above the ground and always projects it at an angle of elevation of at least 35°. If θ is the angle of projection, find the range of possible values of θ that will land the stone in the circle.

Take g as 10 and give angles to the nearest degree.

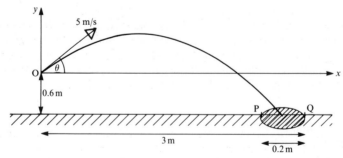

16. A particle is projected from a point O with speed $2\sqrt{gh}$ at an angle α to the horizontal where $\tan \alpha = 2$. Find its height above the point of projection when its horizontal distance from O is

(a) h (b) $2h$ (c) $3h$ (d) $4h$.

Explain the reason for the answer to part (d).

17. A cricketer strikes the ball with speed 36 m s^{-1} at a height of 0.5 m above the pitch. He 'skies' it, hoping to hit it for six. Unfortunately it is caught on the boundary line, 70 m away, at a height of 2.2 m. At what angle was the ball struck?

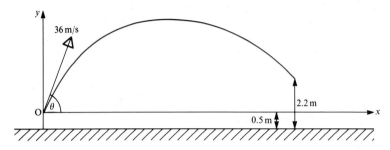

18. The maximum range of a shell fired from a particular gun is 2400 m. At what angle and at what speed is the shell fired?

19. A ball is thrown from O with initial velocity $5\mathbf{i} + 7\mathbf{j}$. Find the Cartesian equation of its path.

20. The speed of a tennis player's serve is 45 m/s. At the moment when the racquet strikes, the ball is exactly over the base line and at a height of 2.8 m. The height of the net is 0.9 m and the distance from the base line to the point where the ball reaches the net is 12 m. If the ball is served at an angle of 8° below the horizontal, will it clear the net? (Take g as 9.81 and work to 3 sf.)

21. A golf ball is struck with a speed of 26 m s^{-1} at an angle α to the horizontal where $\tan \alpha = 2$. The golfer finds that he has not sent the ball in quite the direction he wanted and directly in the line of flight are two trees. One is 22 m high and 20 m from the tee; the other is 21 m high and 48 m from the tee. Determine whether the ball will clear either or both of the trees.

22. A ball is projected with velocity $10\mathbf{i} + 20\mathbf{j}$ from a point O and moves freely under gravity. The ball strikes a wall, 30 m away from O. If $\mathbf{i}$ and $\mathbf{j}$ are vectors of magnitude 1 m s^{-1} horizontally and vertically, respectively, find

(a) the time, in seconds taken for the ball to touch the wall

(b) the height above O, in metres, of the point where the ball strikes the wall

(c) the acute angle, to the nearest degree, which the direction of motion of the ball makes with the horizontal at the instant when it strikes the wall.

FURTHER PROBLEMS

Examples 9c

1. A and B are two points $60\sqrt{3}$ m apart on level ground. A particle P is projected from A towards B with speed 45 m s^{-1} at 30° to the horizontal. At the same instant a particle Q is projected from B towards A with speed $15\sqrt{3}$ m s^{-1} at 60° to the horizontal.

(a) Using exact trig ratios prove that P and Q are always at the same height.

(b) Find t when P and Q collide and find in terms of g their height above the ground when they collide.

For P we will use an x-axis in the direction AB and for Q a separate x-axis in the direction BA.

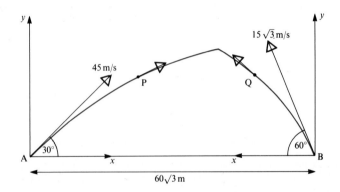

(a) After t seconds, using $y = Vt \sin\theta - \frac{1}{2}gt^2$ gives:

For P $\qquad y_P = 45t \times \frac{1}{2} - \frac{1}{2}gt^2 \qquad = \frac{45}{2}t - \frac{1}{2}gt^2$

For Q $\qquad y_Q = 15\sqrt{3}t \times \frac{\sqrt{3}}{2} - \frac{1}{2}gt^2 = \frac{45}{2}t - \frac{1}{2}gt^2$

i.e. $\qquad$ at any time t, $y_P = y_Q$

∴ $\qquad$ P and Q are always at the same height.

(b) Using $x = Vt \cos\theta$ gives:

For P $\qquad x_P = 45t \times \frac{\sqrt{3}}{2}$

For Q $\qquad x_Q = 15\sqrt{3}t \times \frac{1}{2} = 15t \times \frac{\sqrt{3}}{2}$

P and Q will collide when $\qquad x_P + x_Q = AB$

i.e. when $\qquad\qquad\qquad 60t \times \frac{\sqrt{3}}{2} = 60\sqrt{3}$

$\Rightarrow \qquad\qquad\qquad\qquad t = 2$

When $t = 2 \qquad y = \frac{45}{2}t - \frac{1}{2}gt^2 \qquad \Rightarrow \qquad y = 45 - 2g$

P and Q collide $(45 - 2g)$ m above the ground, when $t = 2$.

2. A particle is projected from a point O on a plane inclined at 30° to the horizontal. The velocity of projection is 20 m s^{-1} at 30° to an upward line of greatest slope. If the particle hits the plane at the point A, find, taking g as 10,

(a) the horizontal distance travelled by the particle

(b) the range up the plane.

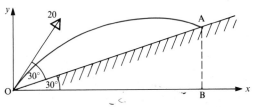

The point A is on the path of the projectile so its coordinates satisfy the equation of the path. A is also on the plane so its coordinates satisfy the equation of the line OA.

The equation of the path of the projectile is

$$y = x \tan 60° - \frac{gx^2}{2(20)^2 \cos^2 60°}$$ [1]

The equation of the line OA is

$$y = x \tan 30°$$ [2]

The coordinates of A can be found by solving these two equations simultaneously.

From [1] and [2]

$$x \tan 30° = x \tan 60° - \frac{10x^2}{800 \cos^2 60°}$$

$$\Rightarrow \qquad \frac{10x}{800 \times 0.25} = 1.7321 - 0.5774$$

$$\Rightarrow \qquad x = 1.1547 \times 20 = 23.094$$

This value of x is also the x coordinate of B.

The particle travels a horizontal distance of 23 m (corrected to 2 sf).

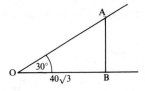

The range up the plane is the distance OA

$$OA = \frac{23.094}{\cos 30°} \text{ m} = 26.6\ldots \text{ m}$$

The range up the plane is 27 m (2 sf).

EXERCISE 9c

The problems in this exercise are varied in type and are generally a little harder than those in the previous exercises. If g is given a numerical value, use 10 and correct answers to 2 significant figures.

1. A particle P is projected from a point O with velocity $12\mathbf{i} + 16\mathbf{j}$. Two seconds later, another particle Q is projected from O and collides with P after another second. Find the initial velocity of Q.

2. For environmental reasons, golfers playing on the village golf course are required to restrict the height of golf balls in flight to 15 m. If a player tees off with an initial speed of 35 m s^{-1} at an angle of projection of 40°, for how long does he contravene the regulations?

3. A particle is projected from a point A on a horizontal plane, with velocity 60 m s^{-1} at 30° to the horizontal. At the same instant a second particle is projected, in the opposite direction with speed 50 m s^{-1}, from a point B on the same plane.

   (a) Given that the particles collide, find the angle of projection of the second particle.

   (b) If the time interval from projection to collision is 1.1 seconds find, to 2 significant figures, the distance between A and B.

4. Two particles A and B are projected at the same instant from the same point O on a horizontal plane. The initial velocity of A is V at 30° to the plane and that of B is $V\sqrt{3}$ at 60° to the plane.

   (a) Show that, as long as both particles are in the air, one of them is vertically above the other.

   (b) Find, in terms of V and g, the distance between the two points where the particles return to the plane.

5. A missile P is projected from a point O with speed 21 m s^{-1} at an angle α to the horizontal. One second later a missile Q is projected from a point 0.3 m below O with initial velocity 31.5 m s^{-1} at an angle β to the horizontal. Given that $\tan \alpha = \frac{4}{3}$ and $\tan \beta = \frac{3}{4}$

   (a) prove that the particles collide

   (b) find the time of the collision

   (c) find the direction in which each missile is moving just before the collision.

6. A particle is projected up a line of greatest slope of a plane inclined at 30° to the horizontal. The initial velocity is 15 m s^{-1} at 30° *to the plane*. Find

   (a) the range up the plane   (b) the time of flight.

7. A particle is projected *down* a line of greatest slope of a plane inclined at 30°
to the horizontal. The initial velocity is 15 m s^{-1} at 60° *to the plane*. Find

(a) the range down the plane   (b) the time of flight.

8. A particle P is projected up a line of greatest slope of a plane inclined at 30°
to the horizontal. The initial velocity of P is inclined at 15° to the plane. If P
strikes the plane after 3 seconds, find the speed of projection.

CHAPTER 10

MOTION IN A PLANE

MOTION WITH VARIABLE ACCELERATION

We saw in Chapter 2 that, when an object is moving in a straight line with constant acceleration, there are a number of standard relationships linking velocity, time, displacement and acceleration.

It is very important to appreciate that these formulae can be used *only for motion with constant acceleration.*

There are many different types of motion in which the acceleration is not constant and we must now investigate ways in which such motion can be analysed.

Rates of Increase

If one quantity, y, depends on another quantity, x, then the rate of increase of y with respect to x is denoted by $\dfrac{\mathrm{d}y}{\mathrm{d}x}$

Now velocity, v, is the rate at which displacement, s, increases with respect to time. So we can write

$$v = \frac{\mathrm{d}s}{\mathrm{d}t}$$

Similarly acceleration, a, is the rate at which velocity increases with respect to time, i.e.

$$a = \frac{\mathrm{d}v}{\mathrm{d}t}$$

These relationships provide a means of solving problems on motion in which displacement, velocity and acceleration can be expressed in terms of time.

DISPLACEMENT, VELOCITY AND ACCELERATION AS FUNCTIONS OF TIME

Suppose that a particle is travelling on a straight line so that its displacement s from a fixed point O on the line after t seconds is given by

$$s = 2t^3 + t^2$$

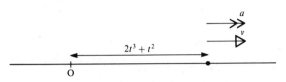

Velocity is the rate of increase of displacement with respect to time, therefore

$$v = \frac{\mathrm{d}s}{\mathrm{d}t} = 6t^2 + 2t$$

i.e. velocity is given by differentiating s with respect to time. Further, acceleration is the rate of increase of velocity with respect to time, therefore

$$a = \frac{\mathrm{d}v}{\mathrm{d}t} = 12t + 2$$

i.e. acceleration is given by differentiating velocity with respect to time. As we found the velocity by differentiating s once, it follows that acceleration is given by differentiating s *twice* with respect to time.

So
$$a = \frac{\mathbf{d^2 s}}{\mathbf{d}t^2}$$

Using the Dot Notation

In Chapter 9, $\dot{x}$ was introduced to indicate differentiating x with respect to time and in a similar way, $\dot{s}$ can be used for $\dfrac{\mathrm{d}s}{\mathrm{d}t}$

Extending this notation, we can use $\ddot{s}$ to represent differentiating s twice with respect to time. So we have

$$s = 2t^3 + t^2 \quad \Rightarrow \quad v = \dot{s} = 6t^2 + 2t$$
$$\Rightarrow \quad a = \dot{v} = \ddot{s} = 12t + 2$$

Hence, for any straight line motion in which the displacement from a fixed point is a function of time, the velocity and the acceleration at any time can be found by differentiation.

Conversely, if we start with the acceleration, then

$$a = \frac{dv}{dt} \qquad \Rightarrow \qquad v = \int a \, dt$$

and

$$v = \frac{ds}{dt} \qquad \Rightarrow \qquad s = \int v \, dt$$

i.e. if the acceleration of a moving body is a function of time,

and

velocity can be found by integrating a with respect to t

displacement can be found by integrating v with respect to t.

Examples 10a

1. A body moves along a straight line so that its displacement s metres, from a fixed point O on the line after t seconds, is given by $s = t^3 - 3t^2 - 9t$

(a) Find the velocity after t seconds.

(b) Find the times when the velocity is zero.

(c) Sketch the velocity–time graph.

(a) $\qquad\qquad s = t^3 - 3t^2 - 9t$

$\qquad \therefore \qquad v = \dot{s} = 3t^2 - 6t - 9$

(b) When $v = 0$, $\qquad 3t^2 - 6t - 9 = 0$

$\qquad \therefore \qquad\qquad 3(t^2 - 2t - 3) = 0 \qquad \Rightarrow \qquad 3(t-3)(t+1) = 0$

$\qquad\qquad\qquad\qquad\qquad\qquad\qquad\qquad \Rightarrow \qquad t = 3 \quad \text{or} \quad -1$

Therefore the velocity is zero *after* 3 seconds; it was also zero 1 second *before* the body reached O.

(c) The expression for the velocity is a quadratic function for which the graph is a parabola crossing the t-axis where $t = 3$ and $t = -1$.

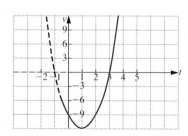

2. A particle P moving in a straight line has an initial velocity of 2 m s^{-1} at a point O on the line. The particle moves so that its acceleration t seconds later is given by $(2t - 6) \text{ m s}^{-2}$. (Remember that m s^{-2} means m/s^2.) Find expressions for the velocity and the displacement of P from O when $t = 5$ and comment on your answers.

$$v = \int a \, dt = \int (2t - 6) \, dt = t^2 - 6t + K_1$$

$v = 2$ when $t = 0$ therefore $K_1 = 2$

$\therefore \qquad v = t^2 - 6t + 2$

When $t = 5$, $\quad v = 25 - 30 + 2 = -3$

The velocity when $t = 5$ is -3 m s^{-1}

i.e. P is moving with speed 3 m s^{-1} *towards* O

$$s = \int v \, dt = \int (t^2 - 6t + 2) \, dt = \tfrac{1}{3} t^3 - 3t^2 + 2t + K_2$$

$s = 0$ when $t = 0$ therefore $K_2 = 0$

$\therefore \qquad s = \tfrac{1}{3} t^3 - 3t^2 + 2t$

When $t = 5$, $\quad s = \tfrac{125}{3} - 75 + 10 = -23\tfrac{1}{3}$

When $t = 5$ the displacement of P from O is $-23\tfrac{1}{3}$ m, i.e. P has changed direction, moving back towards O, and has passed through O to the opposite side.

3. The velocity, v metres per second, of a particle moving in a straight line, is given by $v = (3t^2 - 12t + 9)$ where t is the number of seconds after the particle passes through O, a point on the line. Find

(a) the initial velocity

(b) the time when the acceleration of the particle is zero

(c) the times(s) when the particle passes through O

(d) the times(s) when the direction of the particle is reversed.

(a) The initial velocity occurs when $t = 0$.

  When $t = 0$, $v = 9$

  Therefore the initial velocity is 9 m s^{-1}.

(b) The acceleration at any time is given by $\dfrac{dv}{dt}$, i.e. $a = 6t - 12$

  $a = 0$ when $6t = 12$, i.e. when $t = 2$.

  The acceleration is zero after 2 seconds.

(c)   When the particle passes through O the displacement, s, from O is zero.

$$s = \int v \, dt = \int (3t^2 - 12t + 9) \, dt$$
$$= t^3 - 6t^2 + 9t + K$$

When $t = 0$, $s = 0$,   therefore $K = 0$

$\therefore$ $s = t^3 - 6t^2 + 9t$

The time when the particle passes through O is given by $s = 0$

i.e. $t^3 - 6t^2 + 9t = 0$ $\Rightarrow$ $t(t - 3)^2 = 0$

$\Rightarrow$ $t = 0$ or $t = 3$

The particle is at O initially and after 3 seconds.

(d)   If the direction of motion reverses, the velocity at that instant is zero.

When $v = 0$, $3t^2 - 12t + 9 = 0$ $\Rightarrow$ $3(t^2 - 4t + 3) = 0$

$\Rightarrow$ $3(t - 1)(t - 3) = 0$

$\therefore$ $v = 0$ when $t = 1$ or 3.

The particle's direction is reversed after 1 second and again after another 2 seconds.

4. A particle P is moving on a straight line through a fixed point O. The displacement, s metres, of P from O at time t is given by

$$s = 5 + 9t^2 - 2t^3.$$

Find the *distance* covered in

(a) the first 2 seconds   (b) the first 4 seconds.

Distance and displacement are equal only if the direction of motion does not change within the time interval concerned, so first check whether there are values of t when the direction of motion changes, i.e. when $v = 0$.

$$s = 5 + 9t^2 - 2t^3   \Rightarrow   v = 18t - 6t^2 = 6t(3 - t)$$

$\therefore$ $v = 0$ when $t = 0$ and when $t = 3$

(a)   There is no change in direction of motion between $t = 0$ and $t = 2$ so in the first 2 seconds the distance travelled is equal to the increase in displacement.

When $t = 2$, $s = 5 + 36 - 16$ $\Rightarrow$ $s_2 = 25$

When $t = 0$, $s = 5$ $\Rightarrow$ $s_0 = 5$

The distance travelled in the first 2 seconds is $(s_2 - s_0)$ m

i.e. 20 m.

(b) The direction of motion changes when $t = 3$ so the distance travelled in the first 4 seconds is not equal to the corresponding increase in displacement.

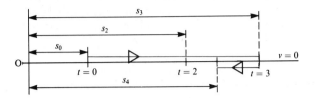

The distance travelled from $t = 0$ to $t = 3$ is $(s_3 - s_0)$ metres.

$$s_3 = 5 + 81 - 54 = 32 \quad \text{and} \quad s_0 = 5 \qquad \Rightarrow \qquad s_3 - s_0 = 27$$

∴   the distance travelled in the first 3 seconds is 27 m.

The distance from $t = 3$ to $t = 4$ is travelled in the opposite direction so is given by $s_3 - s_4$

$$s_3 - s_4 = 32 - (5 + 144 - 128) = 11$$

∴   the distance travelled in the fourth second is 11 m.

The distance travelled in the first 4 seconds is $(27 + 11)$ m

i.e. 38 m.

Variable motion is not always defined by a given function for acceleration, velocity or displacement. Instead we may be given information from which a formula for a relationship can be found.

5. The motion in a straight line of a particle P is such that the acceleration is proportional to $(t + 1)$ at any time t seconds. Initially P has a velocity of 2 m s^{-1} and when $t = 4$ the velocity is 26 m s^{-1}.
Express the velocity and the acceleration as functions of t.

Acceleration $\propto (t + 1)$

i.e. $a = k(t + 1)$ where k is a constant of proportion

$$v = \int a \, dt \quad \Rightarrow \quad v = \int k(t + 1) dt = k(\tfrac{1}{2}t^2 + t) + K$$

When $t = 0$, $v = 2$ $\Rightarrow$ $K = 2$

When $t = 4$, $v = 26$ $\Rightarrow$ $26 = k(8 + 4) + 2$ $\Rightarrow$ $k = 2$

Hence $a = 2(t + 1)$

and $v = t^2 + 2t + 2$

EXERCISE 10a

In each question from 1 to 10, a particle P is moving on a straight line and O is a fixed point on that line. At a time of t seconds the displacement of P from O is s metres, the velocity is v metres per second and the acceleration is a metres per second^2.

1. Given that $s = 4t^3 - 5t^2 + 7t + 6$, find v when $t = 3$.

2. Given that $v = 9t^2 + 14t + 6$, find a when $t = 2$.

3. If $s = t^3 - 2t^2 + 9t$, find a when $t = 5$.

4. Given that P is at O when $t = 0$, and that $v = 2t^2 + 3t + 4$, find s when $t = 4$.

5. P starts from O with velocity 3 m s^{-1}. If $a = 12t - 5$, find v and s when $t = 2$.

6. If $s = t^3 - 9t^2 + 24t - 11$, find the time(s) when $v = 0$.

7. Find the times when the direction of motion of P changes given that $s = 6t^3 - 9t^2 + 4t$.

8. P starts from rest at O and moves with an acceleration given by $2t \text{ m s}^{-2}$. Find v and s in terms of t.

9. When $t = 0$, P passes with a velocity of 4 m s^{-1} through a point with a displacement of 2 m from O. Given that $a = t^2 + 1$, find the velocity and displacement when $t = 4$.

10. When $t = 0$, P passes through O with velocity -4 m s^{-1}. If $a = 8 - 6t$, find

(a) the times when P is instantaneously at rest

(b) the displacement of P from O at these times.

11. A particle P moves in a straight line and O is a fixed point on the line. The displacement, s metres, of P from O at any time t seconds is given by $s = t^3 + t^2 + 12t - 23$. Show that the motion is always in the same direction.

12. A particle moves in a straight line with an acceleration given at any time by $(3t - 1) \text{ m s}^{-2}$. If the particle has a velocity of 3 m s^{-1} and is 7 m from a fixed point O on the line when $t = 2$, find

(a) its velocity when $t = 5$

(b) its displacement from O when $t = 4$.

13. The acceleration of a particle P after t seconds is proportional to $(3t^2 + 1)$. When $t = 3$, the acceleration is 14 m s^{-2} and the speed is 25 m s^{-1}. Find

(a) the acceleration as a function of t (b) the initial velocity.

14. Given that $a = -\dfrac{1}{t^3}$ and that $v = 3$ when $t = 1$,

(a) find the velocity when $t = 4$

(b) show that, as the value of t becomes large, the velocity approaches a particular value (called the terminal velocity) and state this value.

15. A particle starts from rest at a point A and moves along a straight line AB with an acceleration after t seconds given by $a = (8 - 2t^2)$. Find

(a) the greatest speed of the particle in the direction $\overrightarrow{AB}$

(b) the time when this greatest speed occurs

(c) the distance travelled in this time.

16. At any time t, the acceleration of a particle P, travelling in a straight line, is inversely proportional to $(t + 1)^3$. Initially, when $t = 0$, P is at rest at a point O and 3 seconds later it has a speed of 2 m s^{-1}. Find, in terms of t, the displacement of P from O at any time.

17. A particle travelling in a straight line passes initially through a fixed point O on the line with a velocity u. The acceleration of the particle has a constant value a. By using integration find expressions for the velocity v and the displacement s from O, after t seconds.
Compare these results with the standard formulae for motion with constant acceleration.

18. A particle starts from rest at a fixed point A and moves in a straight line with an acceleration which, t seconds after leaving A, is given by $a = 4t$. After 2 seconds the particle reaches a point B and the acceleration then ceases. Find

(a) the velocity when the particle reaches B (b) the distance AB.

Immediately the particle moves on with acceleration given by $-3t$ until it comes to rest at a point C. Find

(c) the value of t when the particle reaches C (d) the distance AC.

19. The displacement, s metres, of a body from a point O after t seconds is given by $s = t^2 + \dfrac{1}{t}$,

(a) Find in terms of t an expression for the acceleration of the body.

(b) Given that the mass of the body is 3 kg, use Newton's Second Law to find the force acting on the body after 5 seconds.

20. An object is moving in a straight line under the action of a force whose value at any time t seconds is given by $F = (12t + 20)$ newtons. When $t = 2$ the object, whose mass is 4 kg, passes through a point A on the line with a velocity of 22 m s^{-1}. Find, as a function of t, the displacement of the object from A at any time.

21. A wagon whose mass is 200 kg is pulled by a cable along a straight level track. Contact between the wagon and the track is smooth and the tension in the cable is directly proportional to the time. The wagon starts from rest and, 10 seconds later, its speed is 20 m s^{-1}. How far has the wagon been pulled?

VARIABLE MOTION IN THE *X-Y* PLANE

When a particle is moving in a plane it can be convenient, as we saw in Chapter 9, to consider separately its motion in two perpendicular directions. If displacement, velocity and acceleration are functions of time, then the calculus methods used above can be applied to the components in each direction.

Consider, for example, the motion of the particle shown in the diagram.

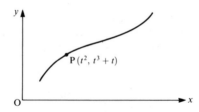

In the direction Ox, at any time t, $x = t^2$

     i.e.   the displacement from O is t^2
     $\therefore$ the velocity is $2t$
     and   the acceleration is 2

In the direction Oy, $y = t^3 + t$

     i.e.   the displacement from O is $t^3 + t$
     $\therefore$ the velocity is $3t^2 + 1$
     and   the acceleration is $6t$

Now we can express these components in terms of unit vectors **i** and **j** in the chosen directions and, by adding them, form a resultant vector.

For example, the displacement vector of P, which is also known as the *position vector* of P, is denoted by **r**, and is given by

$$\mathbf{r} = t^2\mathbf{i} + (t^3 + t)\mathbf{j}$$

Similarly the velocity vector, **v**, is $2t\mathbf{i} + (3t^2 + 1)\mathbf{j}$
and the acceleration vector, **a**, is $2\mathbf{i} + 6t\mathbf{j}$

As each component of **v** is obtained by differentiating the corresponding component of **s** with respect to t, we can say

$$\mathbf{v} = \frac{d\mathbf{s}}{dt} \quad \text{and} \quad \mathbf{s} = \int \mathbf{v} \, dt$$

Similarly
$$\mathbf{a} = \frac{d\mathbf{v}}{dt} \quad \text{and} \quad \mathbf{v} = \int \mathbf{a} \, dt$$

Examples 10b

In these examples, **i** and **j** are perpendicular unit vectors and t is the elapsed time. Acceleration, velocity and displacement are all to be expressed as vectors in **ij** form. All quantities are measured in units based on metres and seconds.

1. A particle is moving in a plane in such a way that its velocity at any time t is given by $2t\mathbf{i} + 3t^2\mathbf{j}$. Initially the position vector of the particle, relative to a fixed point O in the plane, is $5\mathbf{i} - 8\mathbf{j}$. Find, when $t = 3$,

   (a) the acceleration of P   (b) the position vector of P.

(a)
$$\mathbf{v} = 2t\mathbf{i} + 3t^2\mathbf{j}$$
$$\mathbf{a} = \frac{d\mathbf{v}}{dt} = 2\mathbf{i} + 6t\mathbf{j}$$
When $t = 3$, $\mathbf{a} = 2\mathbf{i} + 18\mathbf{j}$

(b)
$$\mathbf{r} = \int \mathbf{v} \, dt = \int (2t\mathbf{i} + 3t^2\mathbf{j}) \, dt$$

When a function is integrated, a constant of integration must be added. In this problem we are integrating a *vector function* so the constant of integration must also be a vector quantity. We will denote it by **A**.

$\therefore$ $\mathbf{r} = t^2\mathbf{i} + t^3\mathbf{j} + \mathbf{A}$

Initially (i.e. when $t = 0$) $5\mathbf{i} - 8\mathbf{j} = 0\mathbf{i} + 0\mathbf{j} + \mathbf{A}$

$\Rightarrow$ $\mathbf{A} = 5\mathbf{i} - 8\mathbf{j}$

$\therefore$ $\mathbf{r} = t^2\mathbf{i} + t^3\mathbf{j} + 5\mathbf{i} - 8\mathbf{j} = (t^2 + 5)\mathbf{i} + (t^3 - 8)\mathbf{j}$

When $t = 3$, $\mathbf{r} = (9 + 5)\mathbf{i} + (27 - 8)\mathbf{j} = 14\mathbf{i} + 19\mathbf{j}$

2. At any time t, the position vector of a particle moving in a plane, relative to a fixed point O in the plane, is $10t\mathbf{i} + (t^4 - 4t)\mathbf{j}$.

(a) Show that the particle has no acceleration in the direction of $\mathbf{i}$.

(b) Find when the velocity is perpendicular to the acceleration.

(c) Find the *distance* from O of the particle when $t = 2$.

(d) Find the angle between the vector $\mathbf{i}$ and the direction of motion when $t = 2$.

(a) $$\mathbf{r} = 10t\mathbf{i} + (t^4 - 4t)\mathbf{j}$$

$$\mathbf{v} = \frac{d\mathbf{r}}{dt} = 10\mathbf{i} + (4t^3 - 4)\mathbf{j}$$

$$\mathbf{a} = \frac{d\mathbf{v}}{dt} = 12t^2\mathbf{j}$$

$\mathbf{a}$ has no term in $\mathbf{i}$, therefore the acceleration has no component in the direction of $\mathbf{i}$.

(b) The acceleration is always in the direction of $\mathbf{j}$. Therefore, in order to be perpendicular to the acceleration, the velocity must be parallel to $\mathbf{i}$.

$$\mathbf{v} = 10\mathbf{i} + (4t^3 - 4)\mathbf{j}$$

$\mathbf{v}$ is perpendicular to $\mathbf{a}$ when the coefficient of $\mathbf{j}$ is zero,

i.e when $4(t^3 - 1) = 0$ $\Rightarrow$ $t = 1$

The velocity is perpendicular to the acceleration after 1 second.

(c) When $t = 2$, $\mathbf{r} = 20\mathbf{i} + (16 - 8)\mathbf{j}$

$$= 20\mathbf{i} + 8\mathbf{j}$$

The distance between O and P is

$$\sqrt{(20^2 + 8^2)}\,\mathrm{m} = 21.5\,\mathrm{m} \quad (3\,\mathrm{sf})$$

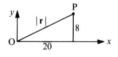

(d) The direction of motion depends upon the components of the velocity.

When $t = 2$, $\mathbf{v} = 10\mathbf{i} + (32 - 4)\mathbf{j}$

$$= 10\mathbf{i} + 28\mathbf{j}$$

$\therefore$ the direction of motion makes an angle α with $\mathbf{i}$,

where $\tan\alpha = \frac{28}{10}$ $\Rightarrow$ $\alpha = 70.3°$ $(3\,\mathrm{sf})$.

3. A force, in newtons, is expressed at any time t seconds by $\mathbf{F} = 2\mathbf{i} + 3(t^2 - 1)\mathbf{j}$. The force acts on a particle P, of mass 2 kg, moving in the xy plane. When $t = 0$, P is at rest at the point with position vector $\mathbf{i} + \mathbf{j}$.

(a) Find (i) the acceleration vector   (ii) the position vector of P at time t.

(b) Write down separate equations for the x and y coordinates of P at time t.

(c) By eliminating t from these two equations, find the Cartesian equation of the path of P and sketch the path.

(a) (i) Using Newton's Law, $\mathbf{F} = m\mathbf{a}$, gives $2\mathbf{i} + 3(t^2 - 1)\mathbf{j} = 2\mathbf{a}$

$$\Rightarrow \quad \mathbf{a} = \mathbf{i} + \tfrac{3}{2}(t^2 - 1)\mathbf{j}$$

(ii) First find the velocity vector.

$$\mathbf{v} = \int \mathbf{a}\, dt = \int [\mathbf{i} + \tfrac{3}{2}(t^2 - 1)\mathbf{j}]\, dt = t\mathbf{i} + (\tfrac{1}{2}t^3 - \tfrac{3}{2}t)\mathbf{j} + \mathbf{A}$$

When $t = 0$, $\mathbf{v} = \mathbf{0}$ therefore $\mathbf{A} = \mathbf{0}$

$$\therefore \quad \mathbf{v} = t\mathbf{i} + (\tfrac{1}{2}t^3 - \tfrac{3}{2}t)\mathbf{j}$$

$$\mathbf{r} = \int \mathbf{v}\, dt = \tfrac{1}{2}t^2\mathbf{i} + (\tfrac{1}{8}t^4 - \tfrac{3}{4}t^2)\mathbf{j} + \mathbf{B}$$

When $t = 0$, $\mathbf{r} = \mathbf{i} + \mathbf{j}$ therefore $\mathbf{i} + \mathbf{j} = \mathbf{B}$

$$\therefore \quad \mathbf{r} = \tfrac{1}{2}t^2\mathbf{i} + (\tfrac{1}{8}t^4 - \tfrac{3}{4}t^2)\mathbf{j} + \mathbf{i} + \mathbf{j}$$

i.e. $\quad \mathbf{r} = (\tfrac{1}{2}t^2 + 1)\mathbf{i} + (\tfrac{1}{8}t^4 - \tfrac{3}{4}t^2 + 1)\mathbf{j}$

(b) $\qquad\qquad x = \tfrac{1}{2}t^2 + 1$ $\qquad\qquad\qquad\qquad$ [1]

$\qquad\qquad y = \tfrac{1}{8}t^4 - \tfrac{3}{4}t^2 + 1$ $\qquad\qquad\qquad$ [2]

(c) From [1] $\quad t^2 = 2(x - 1)$

Substituting in [2] gives $\quad y = \tfrac{1}{8}[4(x - 1)^2] - \tfrac{3}{4}[2(x - 1)] + 1$

Multiplying throughout by 2 gives $\quad 2y = (x - 1)^2 - 3(x - 1) + 2$

i.e. $\quad 2y = x^2 - 5x + 6 = (x - 2)(x - 3)$

This is the Cartesian equation of the path of P which can be recognised as a parabola.

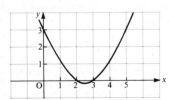

4. At any time t seconds, the position vector of a particle of mass 9 kg is given in metres by $\mathbf{r} = 2t^2\mathbf{i} + 5t\mathbf{j}$. Find the kinetic energy of the particle when (a) $t = 3$ (b) $t = 0$.

The velocity, $\mathbf{v}$, of the particle is given by

$$\mathbf{v} = \frac{d\mathbf{r}}{dt} = 4t\mathbf{i} + 5\mathbf{j}$$

(a) When $t = 3$, $\mathbf{v} = 12\mathbf{i} + 5\mathbf{j}$

   The speed of the particle is given by $|\mathbf{v}|$

   i.e.   the speed is 13 m s^{-1}

   $\therefore$ K.E. $= (\frac{1}{2} \times 9 \times 13^2)$ J $= 760.5$ J

(b) When $t = 0$, $\mathbf{v} = 5\mathbf{j}$ $\Rightarrow$ the speed is 5 m s^{-1}

   $\therefore$ K.E. $= (\frac{1}{2} \times 9 \times 5^2)$ J $= 112.5$ J

EXERCISE 10b

In these questions, a particle P is moving in the xy plane and O is the origin. Unit vectors in the directions Ox and Oy are $\mathbf{i}$ and $\mathbf{j}$ respectively. For P the position vector relative to O at time t is $\mathbf{r}$, the velocity vector is $\mathbf{v}$ and the acceleration vector is $\mathbf{a}$. All units are consistent and based on metres, seconds and newtons.

1. Find, in $\mathbf{ij}$ form, expressions for $\mathbf{v}$ and $\mathbf{a}$
   (i) at any time t (ii) at the specified times,   if

   (a) $\mathbf{r} = 2t^3\mathbf{i} + 3t^2\mathbf{j}$; $t = 2$, $t = 3$

   (b) $\mathbf{r} = t(t+1)\mathbf{i} + (4 - t^2)\mathbf{j}$; $t = 1$, $t = 4$

   (c) $\mathbf{r} = 3t\mathbf{i} + 4t(t^2 + 3)\mathbf{j}$; $t = 1$, $t = 3$

   (d) $\mathbf{r} = \dfrac{2}{t}\mathbf{i} + 3\left(1 + \dfrac{2}{t^2}\right)\mathbf{j}$; $t = 1$, $t = 2$

   (e) $\mathbf{r} = \left(t - \dfrac{1}{t}\right)\mathbf{i} + \left(t^2 - \dfrac{1}{t^2}\right)\mathbf{j}$; $t = 1$, $t = 3$.

2. Find the Cartesian equation of the path traced out by P if

   (a) $\mathbf{r} = 2t\mathbf{i} + 3t^2\mathbf{j}$

   (b) $\mathbf{r} = (t + 1)^2\mathbf{i} + 4t\mathbf{j}$

   (c) $\mathbf{r} = 2t\mathbf{i} + \dfrac{3}{t}\mathbf{j}$

3. The angle between $\mathbf{i}$ and the direction of motion of P is α. Find $\tan\alpha$,
   (i)   at any time t (ii)   at the specified times.

   (a) $\mathbf{r} = 6t\mathbf{i} + (12t - 3t^2)\mathbf{j}$; $t = 0$, $t = 1$, $t = 2$

   (b) $\mathbf{r} = (8t - 10t^2)\mathbf{i} + 6t\mathbf{j}$; $t = 0$, $t = \frac{1}{2}$, $t = \frac{4}{5}$

   (c) $\mathbf{r} = t^2\mathbf{i} + (1 + t^3)\mathbf{j}$; $t = \frac{1}{3}$, $t = \frac{2}{3}$, $t = 1$.

4. Initially P is at rest at a point with position vector $3\mathbf{i} + \mathbf{j}$. Given that the acceleration of P after t seconds is $\mathbf{i} - 2\mathbf{j}$, find in terms of $\mathbf{i}$ and $\mathbf{j}$, expressions for $\mathbf{v}$ and $\mathbf{r}$ at any time t.

5. At any time t, $\mathbf{v} = 3t^2\mathbf{i} + (t - 1)\mathbf{j}$. Given that P is initially at O find in $\mathbf{ij}$ form,

   (a)   the initial velocity

   (b) $\mathbf{a}$ when $t = 3$

   (c) $\mathbf{r}$ when $t = 2$.

6. When $t = 1$, $\mathbf{v} = \mathbf{i} + 3\mathbf{j}$ and $\mathbf{r} = 4\mathbf{i} - \mathbf{j}$. If $\mathbf{a} = t\mathbf{i} + (2 - t)\mathbf{j}$,   find $\mathbf{r}$ when $t = 4$. What is the distance of P from O at this time?

7. The acceleration vector of P is constant and given by $\mathbf{a} = p\mathbf{i} + q\mathbf{j}$. When $t = 0$ the velocity vector is zero and when $t = 1$, $\mathbf{v} = 3\mathbf{i} - 2\mathbf{j}$. Find $\mathbf{v}$ at any time t. What is the *speed* of P when $t = 3$?

8. The coordinates of P at any time t are $(t + t^2,\ 3t^2 - 2)$. Prove that P has a constant acceleration and give its magnitude.

9. A force $\mathbf{F}$ acts on particle of mass 2 kg. Given that $\mathbf{F} = 4t\mathbf{i} + 6\mathbf{j}$,   and that P is initially at O with velocity $5\mathbf{j}$, find $\mathbf{v}$ and $\mathbf{r}$ when $t = 3$.

10. Given that $\mathbf{r} = (2t - 1)\mathbf{i} - t^2\mathbf{j}$,

    (a)   find, in $\mathbf{ij}$ form, the direction of motion
          (i)   initially   (ii)   at time t

    (b)   explain why the direction in which P moves can never be perpendicular to the initial direction of motion

    (c)   show that $\mathbf{a}$ is constant and give its magnitude.

11. Initially P is at O with velocity vector $(V\cos\alpha)\mathbf{i} + (V\sin\alpha)\mathbf{j}$. Given that $\mathbf{a} = -g\mathbf{j}$, where g is the acceleration due to gravity,

    (a)   find, at any time t, expressions for   (i) $\mathbf{v}$ (ii) $\mathbf{r}$

    (b)   hence derive the equation of the path of a particle projected from O with speed V at an angle α to the horizontal.

12. A force represented by $12\mathbf{j}$ is the only force acting on P. Initially P, whose mass is 3 kg, is at O with a speed of 6 m s^{-1} in the direction Ox.

    (a) Express in $\mathbf{ij}$ form   (i) the initial velocity
         (ii) the acceleration vector   (iii) $\mathbf{v}$ and $\mathbf{r}$ after t seconds.

    (b) Show that P moves on a parabola.

13. Two forces $\mathbf{P}$ and $\mathbf{Q}$, of magnitudes P newtons and Q newtons, act on a particle A of mass 2 kg. $\mathbf{P} = 6\mathbf{i} - \mathbf{j}$ and $\mathbf{Q} = 2\mathbf{i} + 5\mathbf{j}$.

    (a) Find the resultant of $\mathbf{P}$ and $\mathbf{Q}$ and hence the acceleration of A.

    (b) Given that A is initially at rest at a point with position vector $3\mathbf{j}$ ($|\mathbf{j}| = 1$ m), find the position vector of P after t seconds.

14. A force $\mathbf{F}_1$, of magnitude 10 N, acts in the direction of $4\mathbf{i} + 3\mathbf{j}$;   the direction of another force $\mathbf{F}_2$, of magnitude $3\sqrt{5}$ N, is $2\mathbf{i} + \mathbf{j}$. Both forces act on a particle P of mass 2 kg, which is initially at rest at the origin. Express in $\mathbf{ij}$ form

    (a) $\mathbf{F}_1$, $\mathbf{F}_2$ and their resultant

    (b) the acceleration of P and the velocity of P after t seconds

    (c) the position vector of P after 2 seconds.

15. A particle P of mass 2 kg is at rest, when a force $4\mathbf{i} + 6t\mathbf{j}$ acts on it for one second. Find, at the end of this time,

    (a) the velocity of P   (b) the kinetic energy of P.

16. At any time t seconds, the position vector of a particle P of mass 4 kg is given by $\mathbf{r} = 6t\mathbf{i} - 4t^2\mathbf{j}$. Only one force, $\mathbf{F}$, acts on P. Find

    (a) the velocity and acceleration vectors of P at time t

    (b) the force $\mathbf{F}$

    (c) the kinetic energy of P after 4 seconds

    (d) the work done by $\mathbf{F}$ in the first 4 seconds.

17. A force represented by $4\mathbf{i} - 6\mathbf{j}$ acts on a particle P of mass 2 kg which is initially at rest at the origin O. Find

    (a) the acceleration of P

    (b) the velocity of P after t seconds

    (c) the kinetic energy of P after 2 seconds.

    During the first 2 seconds, what is

    (d) the work done by the force

    (e) the average power exerted by the force.

RELATIVE VECTORS

Relative Velocity

Most of the time we judge the position or motion of an object with reference to the earth's surface, which is fixed.

Sometimes, however, we 'see' motion that is not relative to the earth. Consider, for example, an observer A sitting in a moving railway carriage, who looks out of the window at B, who is a passenger in another train moving at the same speed on a parallel track. To A, B *appears* to be stationary. Relative to the earth B is moving but, *relative to* A, B is stationary.

Now if A's train is travelling at 70 mph and the speed of B's train is 80 mph, B passes A at 10 mph. The velocities of A and B relative to the earth (often called their true velocities) are 70 mph and 80 mph but, relative to A, the velocity of B is (80–70) mph.

What is happening in both cases is that the observer, in this case A, is disregarding his own velocity and seeing only the *difference* between B's velocity and his own.

Now consider the situation when A gets up from his seat and walks across the carriage to the window opposite, i.e. *relative to the carriage* he has moved at right angles to the railway track.
However, as A crossed the carriage, the carriage itself moved forward along the track; so A's true velocity is at an angle to the track as shown in this diagram.

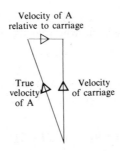

If the true velocities of A and B (i.e. their velocities relative to the earth) are denoted by v_A and v_B and the velocity of B relative to A is denoted by $_Bv_A$ then, in general,

$$_Bv_A = v_B - v_A$$

Examples 10c

The units used throughout are metres and seconds.

1. A passenger on a ship, whose velocity is $14\mathbf{i} - 22\mathbf{j}$, is watching a boat whose velocity is $8\mathbf{i} + 5\mathbf{j}$. What does the velocity of the boat appear to be to the passenger on the ship? Illustrate your solution with a sketch.

The velocity of the ship, $\mathbf{v_S}$, is $14\mathbf{i} - 22\mathbf{j}$

The velocity of the boat, $\mathbf{v_B}$, is $8\mathbf{i} + 5\mathbf{j}$

The velocity of the boat relative to the ship is $_B\mathbf{v_S}$ where

$$_B\mathbf{v_S} = \mathbf{v_B} - \mathbf{v_S}$$
$$= 8\mathbf{i} + 5\mathbf{j} - (14\mathbf{i} - 22\mathbf{j})$$
$$= -6\mathbf{i} + 27\mathbf{j}$$

The boat appears to have a velocity of $-6\mathbf{i} + 27\mathbf{j}$

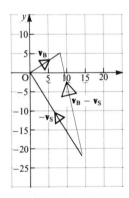

2. The velocity of a particle P is $-2\mathbf{i} + 3\mathbf{j}$. Relative to P, another particle Q has a velocity of $6\mathbf{i} + 9\mathbf{j}$. Find the true velocity of Q and hence find Q's speed.

$$\mathbf{v_P} = -2\mathbf{i} + 3\mathbf{j}$$

$$_Q\mathbf{v_P} = 6\mathbf{i} + 9\mathbf{j}$$

$$\mathbf{v_Q} - \mathbf{v_P} = {_Q\mathbf{v_P}}$$

$$\therefore \qquad \mathbf{v_Q} = {_Q\mathbf{v_P}} + \mathbf{v_P} = 6\mathbf{i} + 9\mathbf{j} + (-2\mathbf{i} + 3\mathbf{j})$$

$$= 4\mathbf{i} + 12\mathbf{j}$$

Q's velocity is $4\mathbf{i} + 12\mathbf{j}$

Q's speed is $|4\mathbf{i} + 12\mathbf{j}|$, i.e. $4\sqrt{10} \text{ m s}^{-1}$.

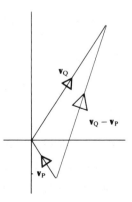

EXERCISE 10c

In questions where units are not specified, any given vectors are measured in units based on metres, seconds, newtons and kilograms.

1. A car, A is travelling at 50 mph on a straight road and another car, B, is being driven on the same road at 40 mph. Find the velocity of B relative to A if the cars are travelling

 (a) in the same direction   (b) in opposite directions.

2. A girl is running at 10 mph after a runaway horse galloping in a straight line at 32 mph.

 (a) What is the velocity of the girl relative to the horse?

 (b) If the horse suddenly stops and begins to gallop at the same speed in the opposite direction, what now is the velocity of the girl relative to the horse?

3. A launch is travelling with velocity $21\mathbf{i} + 16\mathbf{j}$ and the velocity of a pleasure boat is $8\mathbf{i} - 3\mathbf{j}$. Find

 (a) the velocity of the launch relative to the pleasure boat

 (b) the velocity of the pleasure boat relative to the launch.

4. The velocities of a helicopter and a light aircraft are $15\mathbf{i} + 8\mathbf{j}$ and $-3\mathbf{i} + 9\mathbf{j}$. Relative to the light aircraft, what is

 (a) the velocity   (b) the speed   of the helicopter

5. To the pilot of a transport plane flying with velocity $14\mathbf{i} + 11\mathbf{j}$, the velocity of a liner appears to be $-5\mathbf{i} + 2\mathbf{j}$. What is the true velocity of the liner?

6. To a hiker walking due north at 8 km h^{-1} the wind appears to be blowing from the north west at 5 km h^{-1}.

 (a) Taking $\mathbf{i}$ and $\mathbf{j}$ as 1 km h^{-1} east and north respectively, express each of the given velocities in $\mathbf{ij}$ form.

 (b) Find the true wind velocity in the form $a\mathbf{i} + b\mathbf{j}$.

7. The velocity of a particle A is $4\mathbf{i} + 5\mathbf{j}$. Relative to A the velocity of another particle B is $-2\mathbf{i} + \mathbf{j}$ and relative to B the velocity of a third particle C is $3\mathbf{i} - 5\mathbf{j}$. Find

 (a) the true velocity of B

 (b) the true velocity of C

 (c) the velocity of C relative to A.

RELATIVE POSITION

We usually give displacement with respect to a fixed point. For example we could say that two objects, A and B are at points with position vectors $20\mathbf{i} + 16\mathbf{j}$ and $15\mathbf{i} + 22\mathbf{j}$ respectively. These vectors represent the displacements of A and B from the origin and are called the *position vectors* of A and B.

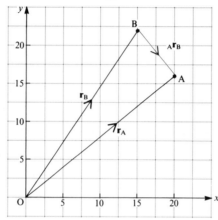

However, from the diagram we see that the displacement of A *relative to B* is $5\mathbf{i} - 6\mathbf{j}$, i.e. $20\mathbf{i} + 16\mathbf{j} - (15\mathbf{i} + 22\mathbf{j})$. This shows that if we use notation similar to the velocity notation, i.e. $\mathbf{r}_A$, $\mathbf{r}_B$ and $_A\mathbf{r}_B$, we have

$$_A\mathbf{r}_B = \mathbf{r}_A - \mathbf{r}_B$$

EXERCISE 10d

In questions 1 to 4 find the displacement of

(a) A relative to B   (b) B relative to A.

1.
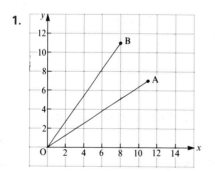

2.
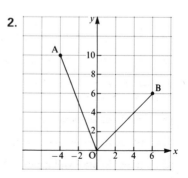

3.
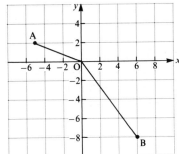

4.
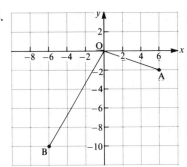

5. Two particles, P and Q, both start moving from the origin with constant velocities $6\mathbf{i} - \mathbf{j}$ and $3\mathbf{i} + 7\mathbf{j}$ respectively.

   (a) Find the position vector of each particle after t seconds.

   (b) What is the displacement of Q relative to P after 3 seconds?

6. A helicopter A leaves a heliport and flies with velocity $10\mathbf{i} + 4\mathbf{j}$. At the same time another helicopter B takes off from a field whose position vector relative to the heliport is $36\mathbf{i} + 2\mathbf{j}$. The velocity of B is $-8\mathbf{i} + 3\mathbf{j}$.

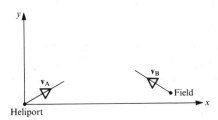

   (a) Find, after t seconds,
       (i)   the position vector of A relative to the heliport
       (ii) the position vector of B relative to the heliport
       (iii) the displacement of B from A.

   (b) Explain what happens when $t = 2$.

7. A particle P starts from rest at the origin when $t = 0$, and moves along the positive x-axis with an acceleration after t seconds given by $\frac{1}{2}t\mathbf{i}$.

   (a) Find the velocity vector and the position vector of P after t seconds.

   A second particle Q starts from the point $(3,0)$ and moves with constant velocity $3\mathbf{i} + 4\mathbf{j}$.

   (b) Find the velocity vector and the displacement vector of Q after t seconds.

   (c) What is the velocity of Q relative to P when $t = 4$?

   (d) What is the displacement of Q relative to P when $t = 4$?

In questions 8 and 9, **i** and **j** are unit vectors east and north respectively. All quantities are measured in units based on kilometres and hours.

8. At 2 p.m. the position vector, relative to a lighthouse, of a ship A is $10\mathbf{i}$ and A's velocity is $12\mathbf{i} + 5\mathbf{j}$. At the same time another ship B, whose velocity is $-3\mathbf{i} + 10\mathbf{j}$, is in a position $20\mathbf{i} - 4\mathbf{j}$ relative to the same lighthouse. Find, after t hours,

(a) the position vector of A relative to the lighthouse

(b) the position vector of B relative to the lighthouse

(c) the position vector of A relative to B

(d) the time when A is due north of B.

9. A particle P starts from the origin O with initial velocity $2\mathbf{i} - \mathbf{j}$ and moves with acceleration $6t\mathbf{j}$. Another particle Q starts from the point **i** with initial velocity $\mathbf{i} + \mathbf{j}$ and moves with acceleration $-4\mathbf{i}$. Find

(a) the velocity of P relative to Q at any time t

(b) the speed of P relative to Q when $t = 2$

(c) the displacement of P relative to Q at any time t

(d) the distance between P and Q when $t = 3$.

CHAPTER 11

IMPULSE AND MOMENTUM

MOMENTUM

The momentum of a body is the product of its mass and its velocity, i.e.

for a body of mass m, moving with velocity v,
momentum $= mv$

Because momentum is a scalar multiple of velocity, which is a vector, it follows that momentum also is a vector quantity.

When the velocity of a body is constant and its mass does not change, its momentum is constant.

We know that a force is needed to change the velocity of an object and it follows that a force must act on the object in order to change its momentum. The precise relationship between a force and the change in momentum that it produces can be found by combining Newton's Second Law with the equations of motion with constant acceleration.

Consider a constant force F that acts for a time t on a body of mass m in the direction of its motion, causing the velocity to increase from u to v. As the force is constant, the acceleration a that it produces, is also constant.

Using $F = ma$ and $v = u + at$ gives

$$v = u + t\left(\frac{F}{m}\right)$$

$$\Rightarrow \quad Ft = mv - mu$$

So we see that the change in momentum, i.e. final momentum minus initial momentum, is given by the product of the force and the time for which it acts.

IMPULSE

The product of the force and the time for which it acts is called the *impulse* of the force and is denoted by the symbol J,

i.e.
$$J = Ft$$

Therefore **impulse = change in momentum**

i.e.
$$J = mv - mu$$

This relationship shows that impulse, too, is a vector quantity. Hence, if a force exerts an impulse on an object in a direction opposite to that of motion, the impulse is negative. It follows that the change in momentum is negative, i.e. the final momentum is less than the initial momentum.

As with all problems involving vector quantities, it is important when dealing with impulse and momentum to define the chosen positive direction.

Although readers will often find that I is used as the symbol for impulse, we prefer to use J. Our reason is that at a further stage in the study of mechanics (beyond the scope of this book) a completely different quantity is represented by I.

Units

The unit of impulse is, as might be expected, the product of a force unit and a time unit so, for a force in newtons acting for a time in seconds, the unit of impulse is the newton second, N s.

(Note that this is *not* newton *per* second.)

Momentum (mass × velocity) can be measured in kilogram metres per second, ($kg\,m\,s^{-1}$) but usually the impulse unit, N s, is used instead.

Examples 11a

1. A hammer of mass 0.8 kg is moving at $12\ m\,s^{-1}$ when it strikes a nail and is brought to rest. What is the magnitude of the impulse exerted on the hammer?

All the initial momentum of the hammer is lost when it hits the nail.

The change in the momentum of the hammer is $0.8 \times 12\ N\,s = 9.6\ N\,s$

∴   the impulse exerted on the hammer is 9.6 N s.

2. A particle of mass 2 kg is moving in a straight line, with a speed of 5 m s^{-1}. A force of 11 N acts on the particle for 6 seconds, in the direction of motion. Find

(a) the magnitude of the impulse exerted on the particle

(b) the speed of the particle at the end of this time.

$$2 \text{ kg} \quad 5 \text{ m s}^{-1} \qquad\qquad 11 \text{ N} \qquad\qquad v \text{ m s}^{-1} \qquad\qquad +\text{ve}$$

Take the positive direction as being to the right.

(a) The impulse of the force is J N s where

$$J = Ft$$

$$= 11 \times 6$$

The magnitude of the impulse is 66 N s.

(b) Initial momentum is 2×5 N s

Final momentum is $2v$ N s

Using $\quad J = mv - mu \quad$ gives

$$66 = 2v - 10 \quad\Rightarrow\quad v = 38$$

The velocity after 6 seconds is 38 m s^{-1}.

3. The velocity of a particle of mass 7 kg, travelling along the x-axis changes from $13\mathbf{i} \text{ m s}^{-1}$ to $3\mathbf{i} \text{ m s}^{-1}$ in 5 seconds under the action of a constant force. Find the magnitude of the force and state its direction.

$$F \text{ newtons}$$

$$13 \text{ m s}^{-1} \qquad\qquad 3 \text{ m s}^{-1}$$

Let the constant force be $\mathbf{F}$ N.

Initial momentum is $7 \times 13\mathbf{i}$ N s; Final momentum is $7 \times 3\mathbf{i}$ N s

The impulse of the force is $5\mathbf{F}$ N s

Using $\quad \mathbf{J} = m\mathbf{v} - m\mathbf{u} \quad$ gives

$$5\mathbf{F} = 21\mathbf{i} - 91\mathbf{i} = -70\mathbf{i}$$

$$\Rightarrow \qquad \mathbf{F} = -70\mathbf{i} \div 5 = -14\mathbf{i}$$

The magnitude of the force is 14 N and it acts in the direction $-\mathbf{i}$.

4. A truck of mass 1200 kg is travelling at 4 m s^{-1} when it hits a buffer and is brought to rest in 3 seconds. What is the average force exerted by the buffer?

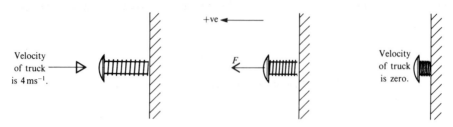

Taking the direction of the force as the positive direction, the initial velocity of the truck is -4 m s^{-1} and the final velocity is zero.

Using $Ft = mv - mu$ gives

$$F \times 3 = 1200 \times 0 - 1200 \times (-4)$$

$\Rightarrow$ $F = 1600$

The average force exerted by the buffer is 1600 N.

5. A particle of mass 3 kg is moving along a straight line in the direction $\overrightarrow{AB}$ with speed 6 m s^{-1} when a force is applied to it. After 4 seconds the particle is moving in the direction $\overrightarrow{BA}$ with speed 2 m s^{-1}. Find the magnitude and direction of the force.

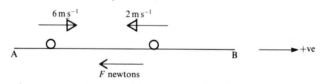

In the direction $\overrightarrow{AB}$

the initial momentum is 3×6 N s,   the final momentum is $3 \times (-2)$ N s and the impulse of the force is $(-F) \times 4$ N s

Using $Ft = mv - mu$ gives

$$-4F = -6 - 18   \Rightarrow   F = 6$$

The force is 6 N acting in the direction $\overrightarrow{BA}$.

Alternatively we could take $\mathbf{i}$ as one unit in the direction $\overrightarrow{AB}$ and use $\mathbf{F}$ as the force vector leading to

$$4\mathbf{F} = 3 \times (-2\mathbf{i}) - 3 \times 6\mathbf{i}   \Rightarrow   4\mathbf{F} = -24\mathbf{i}$$

i.e. $\mathbf{F} = -6\mathbf{i}$

6. A jet of water strikes a wall at right angles to the jet with a speed of 20 m s^{-1}. The water does not bounce off the wall. Given that the average force exerted by the wall in stopping the flow is 360 N, find the mass of water being delivered per second.

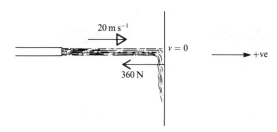

In one second the amount of water that hits the wall is m kilograms.
Therefore the initial momentum of the water is $20m$ N s.
The time taken for m kg of water to be brought to rest is 1 second.

Using $$Ft = mv - mu \quad \text{gives}$$

$$-360 \times 1 = 0 - 20m$$

$\Rightarrow$ $$m = -360 \div (-20) = 18$$

The mass of water delivered per second is 18 kg.

Note that the force which the wall exerts on the water is equal and opposite to the force exerted *by* the water on the wall.

EXERCISE 11a

1. Write down the momentum of

(a)   a child of mass 40 kg running with a speed of 3 m s^{-1}

(b)   a lorry of mass 1200 kg moving at 20 m s^{-1}

(c)   a missile of mass 92 kg travelling at 120 m s^{-1}

(d)   a train of mass 214 tonnes travelling at 55 m s^{-1}

(e)   a bullet of mass 100 g travelling at 40 m s^{-1}.

2. Find the magnitude of the impulse exerted by

(a)   a force of 14 N acting for 6 s

(b)   a force of 12 tonnes acting for 1 minute

(c)   a force that causes an increase in momentum of 88 N s

(d)   the weight of a block of mass 20 kg acting for 30 s.

In questions 3 to 6 a force of magnitude F newtons acts in the direction AB on a particle P of mass 2 kg. Initially the velocity of P is u m s^{-1} and t seconds later it is v m s^{-1}, each in the direction AB.

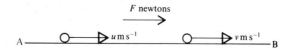

3. If $u = 4$, $v = 7$ and $t = 3$, find F.

4. If $u = 7$, $v = 4$ and $t = 3$, find F.

5. If $u = 5$, $t = 4$ and $F = 10$, find v.

6. If $u = 8$, $t = 5$ and $F = -4$, find v.

7. If $u = 10$, $v = 6$ and $F = -3$, find t.

8. In what time will a force of 12 N reduce the speed of a particle of mass 1.5 kg from 36 m s^{-1} to 12 m s^{-1}?

9. A body of mass 5 kg is moving with a velocity of $10\mathbf{i}$ m s^{-1} when a force $\mathbf{F}$ is applied to it for 4 seconds. Find the velocity at the end of this time if
    (a) $\mathbf{F} = 20\mathbf{i}$ (b) $\mathbf{F} = -20\mathbf{i}$.

10. A body of mass 4 kg is moving with speed 7 m s^{-1} when a force is applied to it for 8 seconds. Its speed then is again 7 m s^{-1} but in the opposite direction. Find the magnitude of the force that has caused this change.

11. A dart of mass 40 g hits the dartboard at a speed of 16 m s^{-1}. If the dart comes to rest in the board in 0.02 seconds, find the average force exerted by the board on the dart.

12. A particle of mass 5 kg has a velocity $16\mathbf{i}$ m s^{-1} when a force $-4\mathbf{i}$ N begins to act on it. Find the velocity of the particle when the force has been acting for (a) $\frac{1}{3}$ s   (b) 5 s.
    After what times will the *speed* of the particle be 2 m s^{-1}?

13. A high pressure hose is being used to clean the wall of a town hall. The hose delivers a horizontal stream of water which hits the wall at a speed of 20 m s^{-1}. Find the average force exerted on the wall, assuming that the water does not bounce back off the wall, if

    (a) 8 kg of water is delivered per second

    (b) the cross-sectional area of the hose pipe is 0.5 cm^2.
        (Take the density of water as 1000 kg m^{-3}.)

14. A stationary truck is shunted into a siding by a locomotive that exerts a force of 2600 N on the truck for 12 seconds.

   (a)   What is the momentum of the truck at the end of this time?

   The truck carries on without change of speed until it is brought to rest in 2 seconds when it hits the buffers at the end of the line.

   (b)   What is the magnitude of the impulse exerted on the truck by the buffers?

   (c)   What is the average force exerted on the truck by the buffers?

IMPULSIVE FORCES

There are circumstances where a large force acts for a very short time, so that neither the force nor the time for which it acts can easily be evaluated separately, e.g. a cricket bat hitting the ball, a shot being fired, a footballer kicking a ball, etc.

These are examples of *impulsive forces* and in such cases the impulse of the force cannot be calculated using $J = Ft$. The change in momentum caused by the impulse, however, can be used to evaluate the impulse.

Example 11b

A cricket ball of mass 0.2 kg has a speed of 20 m s^{-1} when the bat strikes it at right angles and reverses the direction of the ball's flight. If the speed of the ball immediately after being struck is 36 m s^{-1}, find the impulse given by the bat to the ball.

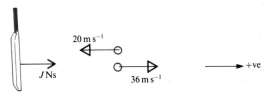

The final momentum of the ball is $0.2 \times 36 \text{ N s} = 7.2 \text{ N s}$
The initial momentum of the ball is $0.2 \times (-20) \text{ N s} = -4 \text{ N s}$
The impulse, J N s, given by the bat is given by

$$J = 7.2 - (-4) = 11.2$$

Therefore the bat exerts an impulse of 11.2 N s on the ball.

EXERCISE 11b

In each question calculate the impulse given.

1. A ball of mass 1.1 kg strikes a wall at right angles with a speed of 6 m s^{-1} and bounces off at 5 m s^{-1}.

$$J = 1.1 \times 5 - (-6 \times 1.1).$$
$$J = 12.1 \text{ Ns}$$

2. The speed of a tennis ball just before it hits the racquet is 38 m s^{-1}. The racquet strikes the ball at right angles, giving it a return speed of 30 m s^{-1}. The mass of the ball is 0.15 kg.

$$J = 0.15 \times 30 - (-38 \times 0.15) = 10.2 Ns.$$

3. A shot of mass 50 g, fired at 250 m s^{-1}, is stopped dead when it hits a steel barrier.

$$J = 0.05 \times 250 - 0$$
$$= 12.5 Ns.$$

4. A bird of mass 60 g is stunned when it flies at 12 m s^{-1} directly into a window pane.

$$m = 60g. \qquad u = 0 \qquad J = 0.06 \times 12 - 0$$
$$m = 0.06 kg \qquad V = 12 ms^{-1} \qquad J = 0.72 Ns$$

5. A stone, of weight 24 N, dropped from a high window, hits the ground at 45 m s^{-1} and does not bounce. (Take g as 10.)

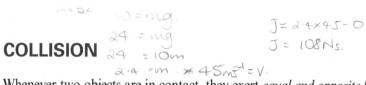

$$m = 24 \qquad W = mg.$$
$$24 = mg$$

COLLISION
$$24 = 10m$$
$$2.4 = m \qquad 45 ms^{-1} = V.$$

$$J = 2.4 \times 45 - 0$$
$$J = 108 Ns.$$

Whenever two objects are in contact, they exert *equal and opposite* forces on each other.

Whether they are in contact for a measurable time, or just for a split second, it is clear that each is in contact with the other for the *same time*. Therefore they exert *equal and opposite impulses* on each other.

Provided that neither object is fixed, these equal and opposite impulses produce equal and opposite changes in momentum, so the overall change in momentum of the two objects caused by the collision is zero. Hence, so long as no external force acts on either object, the total momentum of the two objects (which we refer to as *the system*) remains constant.

This property is known as The Principle of Conservation of Linear Momentum and is expressed formally as follows:

If in a specified direction, no external force affects the motion of a system, the total momentum in that direction remains constant.

In some problems involving a collision the two colliding objects bounce and so have individual velocities after impact. Other objects collide and join together at impact, e.g. trucks which become coupled. Such objects are said to *coalesce*.

In either case we are dealing with different velocities before and after impact so it is advisable to draw separate 'before' and 'after' diagrams and to define the chosen positive direction particularly carefully.

JERK IN A STRING

Consider two particles A and B connected by an inextensible string of length ℓ and lying on a smooth table. The distance between A and B is less than ℓ, so the string is not taut.

Now if A is projected away from B it will move with constant velocity until AB $= \ell$. At that instant the string jerks tight and suddenly exerts equal and opposite tensions on A and B.

These impulsive tensions cause A and B to experience equal and opposite changes in momentum. Therefore, just as in the case of collision, the total momentum of the system is unchanged by the jerk, i.e. the principle of conservation of linear momentum can be applied.

Now the impulse that acts on B jerks B into motion, while the impulse that acts on A gives A a jerk backwards and, because the string is now taut, A and B begin to move on with equal speeds.

Examples 11c

1. A particle A of mass 3 kg, travelling at 5 m s^{-1} collides head-on with a particle B with mass 2 kg and travelling at 4 m s^{-1}. If, after impact, B moves in the opposite direction at 2 m s^{-1}, find the velocity of A.

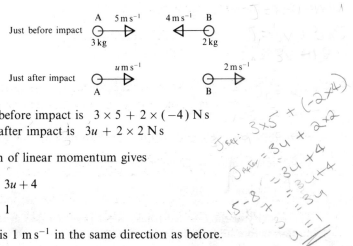

Total momentum before impact is $3 \times 5 + 2 \times (-4)$ N s
Total momentum after impact is $3u + 2 \times 2$ N s

Using conservation of linear momentum gives

$$15 - 8 = 3u + 4$$

$\Rightarrow \qquad u = 1$

The velocity of A is 1 m s^{-1} in the same direction as before.

Note that in this example the direction of motion of A after collision is not given. We guessed that it was to the right and got a positive value for u showing that the guess was correct. Had we thought that A moved with speed u to the left, we would have found that $u = -1$; this also shows that A moves to the right. In other words it does not matter in which direction we mark an unknown velocity; the sign of the answer defines the correct direction.

2. A three-tonne truck is moving along a track at 8 m s^{-1} towards a five-tonne truck travelling at 5 m s^{-1} on the same track. If the trucks become coupled at impact find the velocity at which they continue to move if they are travelling

(a)   in the same direction   (b)   in opposite directions.

(a)

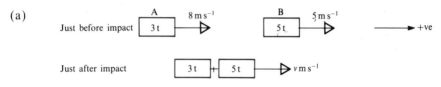

Before impact,   the momentum of A is $3000 \times 8 \text{ N s} = 24\,000 \text{ N s}$
                 the momentum of B is $5000 \times 5 \text{ N s} = 25\,000 \text{ N s}$ $+$
                 the total momentum is $49\,000 \text{ N s}$

After impact,   the combined momentum is $8000 \times v \text{ N s}$

Using conservation of linear momentum gives

$$8000v = 49\,000 \qquad \Rightarrow \qquad v = 6.125$$

The velocity of the coupled trucks is 6.1 m s^{-1} (1 dp).

(b)

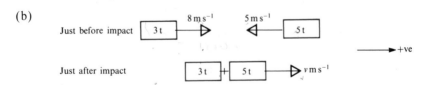

Before impact,   the momentum of A is $3000 \times 8 \text{ N s} = 24\,000 \text{ N s}$
                 the momentum of B is $5000 \times (-5) \text{ N s} = -25\,000 \text{ N s}$
                 the total momentum is -1000 N s

After impact,   the combined momentum is $8000 \times v \text{ N s}$

Using conservation of linear momentum gives

$$8000v = -1000 \qquad \Rightarrow \qquad v = -0.125$$

The velocity of the coupled trucks is 6.1 m s^{-1} (1 dp) in the direction of motion before impact of the heavier truck.

3. Two particles A and B, joined by a light inextensible string, are lying together on a smooth horizontal plane. The masses of A and B are 1 kg and 1.5 kg respectively. A is projected away from B with a speed of 5 m s⁻¹. Find the speed of each particle after the string jerks taut.

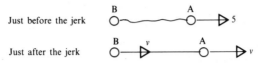

Just before the jerk

Just after the jerk

The momentum of the system before the jerk is 1×5 N s

When the string becomes taut, its ends have equal speeds, i.e. A and B have equal speeds.

The momentum of the system after the jerk is $(1 + 1.5) \times v$ N s

Conservation of linear momentum gives

$$5 = 2.5v \quad \Rightarrow \quad v = 2$$

Each particle has a speed of 2 m s⁻¹ after the string jerks taut.

EXERCISE 11c

Keep your solutions to the questions in this exercise; you will need to refer to them in a later exercise.

In each question from 1 to 5 a body A of mass m_A travelling with velocity u_A, collides directly with a body B of mass m_B moving with velocity u_B. They coalesce at impact. The velocity with which the combined body moves on is v.

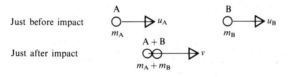

Just before impact

Just after impact

1. $m_A = 4$ kg, $u_A = 4$ m s⁻¹, $m_B = 2$ kg, $u_B = 1$ m s⁻¹. Find v.

2. $m_A = 6$ kg, $u_A = 1$ m s⁻¹, $m_B = 2$ kg, $u_B = -3$ m s⁻¹. Find v.

3. $m_A = 9$ kg, $u_A = 5$ m s⁻¹, $m_B = 4$ kg, $v = 3$ m s⁻¹. Find u_B.

4. $m_A = 3$ kg, $u_A = 16$ m s⁻¹, $m_B = 5$ kg, $v = 6$ m s⁻¹. Find u_B.

5. $u_A = 3$ m s⁻¹, $m_B = 6$ kg, $u_B = -5$ m s⁻¹, $v = -1$ m s⁻¹. Find m_A.

6. A bullet of mass 0.1 kg is fired horizontally, at 80 m s⁻¹, into a stationary block of wood that is free to move on a smooth horizontal plane. The wooden block, with the bullet embedded in it, moves off with speed 5 m s⁻¹. Find the mass of the block.

7. A particle A of mass 5 kg travelling with speed $6 \ \mathrm{m\,s^{-1}}$, collides directly with a stationary particle B of mass 10 kg. If A is brought to rest by the impact find the speed with which B begins to move.

8. The masses of two particles, P and Q, are respectively 0.18 kg and 0.1 kg. They are moving directly towards each other at speeds of $4 \ \mathrm{m\,s^{-1}}$ and $12 \ \mathrm{m\,s^{-1}}$ respectively. After they collide the direction of motion of each particle is reversed and the speed of Q is $6 \ \mathrm{m\,s^{-1}}$. Find P's speed after impact.

9. A sphere P of mass 2 kg is moving at $4 \ \mathrm{m\,s^{-1}}$ when it collides with a sphere Q of mass 1 kg moving in the same direction at $3 \ \mathrm{m\,s^{-1}}$. After the impact, both P and Q move on in the same direction as before, P at $u \ \mathrm{m\,s^{-1}}$ and Q at $v \ \mathrm{m\,s^{-1}}$. Given that $7u = 2v$, find u and v.

In this diagram, A and B are particles resting on a smooth table and connected by a slack light inextensible string. Use the diagram for questions 10 and 11.

10. A is of mass 2 kg and B is of mass 4 kg. B is projected with a speed of $10 \ \mathrm{m\,s^{-1}}$. Find the common speed of A and B when the string jerks taut.

11. B is of mass 0.5 kg and is projected with speed $12 \ \mathrm{m\,s^{-1}}$. The jerk when the string becomes taut causes both particles to begin to move with a speed of $4 \ \mathrm{m\,s^{-1}}$. Find the mass of A.

12. Two particles A and B of equal mass m are connected by a light inextensible string of length ℓ. Initially they are held at rest, side by side. A is then released from rest.

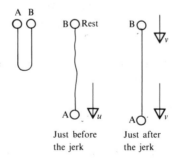

(a) Find, in terms of ℓ and g, A's speed just as the string is about to jerk taut.

(b) If B is released at this instant find, in terms of ℓ and g, the common speed with which A and B together begin to move.

LOSS IN KINETIC ENERGY

When there is a sudden change in the motion of a system, there is usually a change in the total kinetic energy of the system.
Consider the system given in question 1 in Exercise 11c.

Just before impact   4 kg   O—▷ 4 m s⁻¹   2 kg   O—▷ 1 m s⁻¹

Just after impact   6 kg   OO—————▷ 3 m s⁻¹

You should have found the common speed after impact to be 3 m s^{-1}
Before impact the KE of A is $\frac{1}{2}(4)(4^2)$ J and that of B is $\frac{1}{2}(2)(1^2)$ J.
Therefore the total KE of the system is $(32+1)$ J, i.e. 33 J

After impact the KE of A is $\frac{1}{2}(4)(3^2)$ J and that of B is $\frac{1}{2}(2)(3^2)$ J.
Therefore the total KE of the system is $(18+9)$ J, i.e. 27 J

From this we see that a loss in KE of 6 J has been caused by the collision.
This can be understood by remembering that when objects collide there is usually a 'bang', i.e. some mechanical energy is converted into sound energy; some may also be converted into heat energy.

FINDING THE IMPULSE

We know that when objects collide, equal and opposite impulses act on the two objects. When we consider the system as a whole, these impulses cancel and do not appear in our calculations.

If the magnitude of the impulses is required we must consider *only one* of the colliding objects. The impulse which acts on one object causes the change in momentum of that object alone.

In the same way the impulsive tension in a jerking string can be found by considering its effect on the particle at one end only.

Looking again at question 1 in Exercise 11c:

Just before impact   A   O—▷ 4 m s⁻¹   B   O—▷ 1 m s⁻¹
     4 kg   2 kg

At impact J Ns ◁———— A B OO ————▷ J Ns

Just after impact   A + B OO—————▷ 3 m s⁻¹

For the body A alone: the initial momentum is 4×4 N s
the final momentum is 4×3 N s

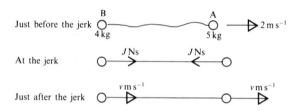

Using   Impulse $=$ Change in momentum   gives

$$-J = 12 - 16 \qquad \Rightarrow \qquad J = 4$$

The impulse that acts on A is 4 N s.
(The impulse that acts on B is also 4 N s but in the opposite direction.)

To find the impulse in a jerking string (often called an impulsive tension) a very similar approach is used.
Suppose, for example, that a mass of 5 kg is projected at speed 2 m s^{-1} away from a mass of 4 kg to which it is attached by a slack inelastic string.

Just before the jerk   B $\underset{4\,kg}{\bigcirc}\!-\!\!-\!\!-\!\!-\!\!-\!\!-\!\!-\!\!-\!\!-\!\!-\!\!-\!\!\underset{5\,kg}{\overset{A}{\bigcirc}} \rhd 2\,m\,s^{-1}$

At the jerk $\bigcirc \overset{J\,Ns}{\longrightarrow} \qquad \overset{J\,Ns}{\longleftarrow} \bigcirc$

Just after the jerk $\bigcirc \overset{v\,m\,s^{-1}}{\rhd} \!\!-\!\!-\!\!-\!\!-\!\!-\!\!-\!\!-\!\!-\!\! \bigcirc \overset{v\,m\,s^{-1}}{\rhd}$

First we need the common speed, v m s^{-1}, after the string jerks tight. This is found from conservation of linear momentum,

i.e. $5 \times 2 = (5+4)v \qquad \Rightarrow \qquad v = \frac{10}{9}$

Now for B alone we use   Impulse $=$ Change in momentum   giving

B
$\bigcirc$ At rest

$\bigcirc\!\!\longrightarrow J\,Ns$

$\bigcirc\!\!-\!\!\rhd \frac{10}{9}\,m\,s^{-1}$

$J =$ final momentum $-$ initial momentum

$= 4 \times \left(\frac{10}{9} \right) - 0$

$\therefore$ the impulse in the string is $4\frac{4}{9}$ N s.

EXERCISE 11d

1. For parts (a), (b), (c), (d), (e) and (f), use your solutions to questions 2 to 5, 7 and 8 in Exercise 11c, to find in each case
    (i) the loss in kinetic energy caused by the impact
    (ii) the impulse exerted on each object.

2. Using your solutions to questions 10 and 11 in Exercise 11c, find in each case the impulse in the string when it jerks tight.

3. A truck A of mass 400 kg, moving at 2 m s^{-1}, runs into a stationary truck B. The two trucks become coupled together and move on with speed 0.8 m s^{-1}. Find

   (a)   the mass of truck B

   (b)   the impulse exerted on truck B by truck A

   (c)   the loss in kinetic energy caused by the collision.

4. The masses of two particles A and B are m and $3m$ respectively. They lie at rest on a smooth horizontal plane and are joined by a light inextensible string.

   (a)   A is projected directly away from B with speed $4u$.

   (b)   B is projected directly away from A with speed $4u$.

   In each case find

    (i)   the common speed of A and B after the string jerks tight

   (ii)   the impulsive tension in the string

   (iii) the loss in kinetic energy caused by the jerk in the string.

5. A sphere of mass 3 kg is dropped on to a horizontal plane from a height of 2 m above the plane. Take g as 10.

   (a)   Find the speed of the sphere when it hits the plane.

   (b)   If the sphere does not bounce find the impulse it exerts on the plane.

   If the sphere rebounds to a height of 1.4 m, find

   (c)   the speed at which the sphere rises off the plane

   (d)   the impulse exerted by the plane on the sphere.

CONSOLIDATION C

SUMMARY

Projectiles

For a particle moving in a vertical xy plane with initial speed V at an angle θ to the horizontal, at any time t:

$$\dot{x} = V \cos \theta \qquad \dot{y} = V \sin \theta - gt$$

$$x = Vt \cos \theta \qquad y = Vt \sin \theta - \tfrac{1}{2}gt^2$$

The equation of the path is

$$y = x \tan \theta - \frac{gx^2}{2V^2 \cos^2 \theta} \quad \text{or} \quad y = x \tan \theta - \frac{gx^2}{2V^2}(1 + \tan^2 \theta)$$

The greatest height occurs when $\dot{y} = 0$

The range on a horizontal plane is $\dfrac{V^2 \sin 2\theta}{g}$ and is found by using $y = 0$

The maximum range is $\dfrac{V^2}{g}$ and is given when $\theta = 45°$

Variable Motion

If displacement is a function of time then

$$v = \frac{ds}{dt} \quad \text{and} \quad a = \frac{dv}{dt}$$

Also $\quad s = \int v \, dt \quad \text{and} \quad v = \int a \, dt$

This applies equally well when $\mathbf{a}$, $\mathbf{v}$ or $\mathbf{s}$ are vectors given in $\mathbf{i}\,\mathbf{j}$ form.

For example, $\quad \mathbf{a} = 2t\mathbf{i} + 3t^2\mathbf{j}$

$\Rightarrow \qquad\qquad \mathbf{v} = t^2\mathbf{i} + t^3\mathbf{j} + \mathbf{K}$ where $\mathbf{K}$ is a *vector*

Relative Motion

If an object A is at a position vector s_A with velocity v_A and another object B is at a position vector s_B with velocity v_B then

the displacement of A relative to B is $s_A - s_B$
and the velocity of A relative to B is $v_A - v_B$

Momentum and Impulse

The momentum of a body of mass m and velocity v is mv.

The impulse of a force F acting for a time t is Ft.

Impulse $=$ Change in momentum

Impulse and momentum are measured in the same unit, the newton second, N s.

At the instant of a collision or a jerk, an instantaneous impulse occurs. The value of an instantaneous impulse can be found only from the change in momentum it produces.

MISCELLANEOUS EXERCISE C

In this exercise use $g = 9.8$ unless another instruction is given.

1. A brick of mass 3 kg falls from rest at a vertical height of 8 m above firm horizontal ground. It does not rebound.
   Calculate the impulse of the force exerted on the ground by the brick.   (AEB)

2. A particle of mass 3 kg moves under the action of a force F N. At time t seconds the velocity v m s^{-1} of the particle is given by $v = 3i + 2tj$.

   (a) Find F.

   (b) Find the kinetic energy of the particle at time t seconds.

   (c) Given that the position vector of the particle when $t = 0$ is $i + j$, find its position vector when $t = 2$.   (NEAB)

3. A body of mass 2 kg, moving along a straight line with speed 5 m s^{-1}, collides with a body of mass 1 kg moving in the same direction along the same straight line with speed 2 m s^{-1}. On collision the bodies adhere and move on together. Calculate

   (a) their common speed immediately after the collision

   (b) the kinetic energy lost during the collision.   (NEAB)

4. The position vector, with respect to a fixed origin O, of a particle at time t
   seconds is $\mathbf{r}$ metres, where $\mathbf{r} = 10t\mathbf{i} + (t^2 - 3)\mathbf{j}$.

   (a) Find the velocity vector of the particle at time t seconds.

   (b) Find, in m s^{-1} to 1 dp, the speed of the particle when $t = 2$.   (ULEAC)

5. A particle is projected from a point on horizontal ground. The initial
   horizontal and vertical components of velocity are 14 m s^{-1} and 21 m s^{-1}
   respectively. Find

   (a) the times when the particle is moving at 45° to the horizontal

   (b) the horizontal and vertical distances from the point of projection to the
   particle at *each* of these times.   (WJEC)

6. A cricket ball, of mass 0.14 kg, is moving horizontally with speed 27 m s^{-1}
   when it hits a vertically held cricket bat. The ball rebounds horizontally with
   speed 15 m s^{-1}.
   Calculate the magnitude of the impulse, in N s, of the force exerted by the ball on
   the bat.   (ULEAC)

7. A stone is thrown, at an angle of 30° above the horizontal, from the edge of a
   vertical cliff at a height of 35 m above sea level. If the initial speed of the stone
   is 7 m s^{-1}, find the time taken for the stone to hit the sea. Find also the
   horizontal distance from the bottom of the cliff at which the stone enters the sea.
                                                                                (AEB)

   Each question from 8 to 11 is followed by several suggested responses. Choose
   which is the correct response.

8. A projectile is projected from a point O on level ground with initial velocity u
   at 45° to the horizontal. When it is about to hit the ground

   A $y = 0$ B $x = 0$ C   it is travelling vertically downwards.

9. A particle of mass 3 kg moves along a straight line Ox under the action of a
   force F such that a time t, $x = t^2 + 3t$. The magnitude of F at time t is given
   by

   A   0   B   5 N   C $3(2t + 3)$ D   6 N

10. Two masses collide and coalesce as shown in the diagram. What is the speed
    V of the combined mass just after impact?

   A $3v$ B $\frac{3}{5}v$ C v D $\frac{5}{3}v$.

11. A particle of mass 2 kg moving with speed 4 m s^{-1} is given a blow which changes the speed to 1 ms^{-1} without deflecting the particle from a straight line. The impulse of the blow is:

   A   10 N s

   B   6 N s

   C   we do not know whether it is 10 N s or 6 N s.

12. With respect to a fixed origin O, the position vector, **r** metres, of a particle P, of mass 0.15 kg, at time t seconds is given by

   $$\mathbf{r} = t^3\mathbf{i} - 4t^2\mathbf{j}, \qquad t \geqslant 0.$$

   (a)   Find, in m s^{-1}, the speed of P when $t = 2$.

   (b)   Find, in N to 2 decimal places, the magnitude of the resultant force acting on P when $t = 2$.   (ULEAC)

13. A quarry railway truck, of mass 1500 kg, is moving at a speed of 6 m s^{-1} along a horizontal track. It collides with a stationary empty truck, of mass 500 kg. The two trucks immediately couple together and move on together. Calculate

   (a)   the speed, in m s^{-1} of the pair of trucks immediately after the collision

   (b)   the total loss of kinetic energy, in J, due to the collision

   (c)   the magnitude of the impulse, in N s, on the stationary truck due to the collision.   (ULEAC)

14.

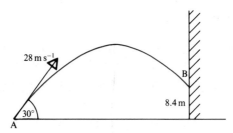

A golf ball is driven from a point A with a velocity which is of magnitude 28 m s^{-1} and at an angle of elevation of 30°. The ball moves freely under gravity. On its downward flight, the ball hits a vertical wall, at a point B, which is 8.4 m above the level of A, as shown in the diagram. Calculate

   (a)   the greatest height achieved by the ball above the level of A,

   (b)   the time taken by the ball to reach B from A .

By using the principle of conservation of energy, or otherwise,

   (c)   find the speed, in m s^{-1} to 1 decimal place, with which the ball strikes the wall.   (ULEAC)

15. The unit vectors **i** and **j** are horizontal and vertically upwards respectively. A particle is projected with velocity $(8\mathbf{i} + 10\mathbf{j})\ \text{m s}^{-1}$ from a point O at the top of a cliff and moves freely under gravity. Six seconds after projection the particle strikes the sea at the point S.
    Calculate

    (a) the horizontal distance between O and S

    (b) the vertical distance between O and S to the nearest metre,

    (c) the speed with which the particle strikes the sea, giving your answer in m s^{-1} to 1 decimal place.   (ULEAC)

16. A particle P moves along the x-axis passing through the origin O at time $t = 0$. At any subsequent time t seconds, P is moving with a velocity of magnitude $v\ \text{m s}^{-1}$ in the direction of x increasing where

    $$v = 2t^3 + 2t + 3, \qquad t \geqslant 0.$$

    (a) Find the acceleration of P when $t = 3$.

    (b) Find the distance covered by P between $t = 0$ and $t = 4$.

    A second particle Q leaves O when $t = 1$ with constant velocity of magnitude $10\ \text{m s}^{-1}$ in the direction of the vector $3\mathbf{i} - 4\mathbf{j}$, where **i** and **j** are unit vectors parallel to Ox and Oy respectively.

    Find, as a vector in terms of **i** and **j**,

    (c) the velocity of Q

    (d) the velocity of P relative to Q at the instant when $t = 1$.

    Hence

    (e) find the magnitude of the velocity of P relative to Q when $t = 1$

    (f) find the angle between the relative velocity and the vector **i** at this instant.
                                                                    (ULEAC)

17. With respect to a fixed origin O, the unit vectors **i** and **j** are directed horizontally and vertically upwards respectively. At time $t = 0$ a ball is projected with velocity $(10\mathbf{i} + 20\mathbf{j})\ \text{m s}^{-1}$ from O and moves freely under gravity. The ball strikes a vertical post, which is 30 m horizontally away from O, at a point P above the horizontal plane through O. Calculate

    (a) the time t, in seconds, when the ball strikes the post

    (b) the height, in metres to 2 significant figures, of P above the plane

    (c) the acute angle, in radians to 2 significant figures, which the velocity of the ball makes with the horizontal at the instant when it strikes the post.
                                                                    (ULEAC)

18.

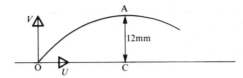

A missile P is projected from a point O on horizontal ground with a velocity whose components are U horizontally and V vertically upwards. At the point A on its flight path the missile is at its maximum height H above the ground. The time taken for the missile to travel from O to A is T. Express both H and T in terms of V and g.

When P has been travelling for a time $\frac{2}{3}T$, a second missile Q is projected vertically upwards with speed V_1 from the point C which is on the ground vertically below A. If Q subsequently collides with P at A, show that $V_1 = \frac{5}{3}V$.

Given also that $OC = \dfrac{3U^2}{8g}$, find V in terms of U.

Show that, just before P and Q collide, the speed of P is twice that of Q.

(NEAB)

19. The unit vectors **i** and **j** are directed due east and due north respectively. The airport B is due north of airport A. On a particular day the velocity of the wind is $(70\mathbf{i} + 25\mathbf{j})\,\text{km}\,\text{h}^{-1}$. Relative to the air an aircraft flies with constant speed $250\,\text{km}\,\text{h}^{-1}$. When the aircraft flies directly from A to B determine.

(a) its speed, in $\text{km}\,\text{h}^{-1}$, relative to the ground

(b) the direction, to the nearest degree, in which it must head.

After flying from A to B, the aircraft returns directly to A.

(c) Calculate the ratio of the time taken on the outward flight to the time taken on the return flight. (ULEAC)

20.

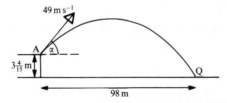

A golf ball is projected with speed $49\,\text{m}\,\text{s}^{-1}$ at an angle of elevation α from a point A on the first floor of a golf driving range. Point A is at a height of $3\frac{4}{15}$ m above horizontal ground. The ball first strikes the ground at a point Q which is at a horizontal distance of 98 m from the point A as shown in the diagram.

(a) Show that $6\tan^2\alpha - 30\tan\alpha + 5 = 0$

(b) Hence find, to the nearest degree, the two possible angles of elevation.

(c) Find, to the nearest second, the smallest possible time of direct flight from A to Q. (ULEAC)

21. A river with long straight banks is 500 m wide and flows with a constant speed of 3 m s^{-1}. A man rowing a boat at a steady speed of 5 m s^{-1}, relative to the river, sets off from a point A on one bank so as to arrive at the point B directly opposite A on the other bank. Find the time taken to cross the river.

    A woman also sets off at A rowing at 5 m s^{-1} relative to the river and crosses in the shortest possible time. Find this time and the distance downstream of B of the point at which she lands. (AEB)

22. The two forces $(4\mathbf{i} - 6\mathbf{j})$ N and $(6\mathbf{i} + 2\mathbf{j})$ N act on a particle P, of mass 2 kg.

    (a) Show that the acceleration of P is $(5\mathbf{i} - 2\mathbf{j})$ m s^{-2}.

    At time $t = 0$, P is at a point with position vector $(5\mathbf{i} + 3\mathbf{j})$ m relative to a fixed origin O and has velocity $(-7\mathbf{i})$ m s^{-1}. Calculate, at time $t = 2$ seconds,

    (b) the position vector of P relative to O

    (c) the velocity, in m s^{-1}, of P

    (d) the kinetic energy, in J, of P. (ULEAC)

23. While practising her tennis serve, Jenny hits the ball from a height of 2.5 m with a velocity $(25\mathbf{i} - 0.5\mathbf{j})$ m s^{-1}. ($\mathbf{i}, \mathbf{j}$ represent unit vectors in horizontal and vertical directions respectively.)

    (a) Find the horizontal distance from the serving point at which the ball lands.

    (b) Determine whether the ball would clear a net, which is 1 m high and 12 m from her serving position in the horizontal direction, $\mathbf{i}$. (AEB)

24. Sally can swim in still water at 1.1 m s^{-1}. She swims across a river flowing at 0.7 m s^{-1} between parallel banks 25 m apart. Find the time, in seconds to 1 decimal place, she takes to swim from a given point on one bank to the nearest point on the opposite bank. (ULEAC)

25. A particle P moves from rest, at a point O at time $t = 0$ seconds, along a straight line. At any subsequent time t seconds the acceleration of P is proportional to $(7 - t^2)$ m s^{-2} and the displacement of P from O is s metres. The speed of P is 6 m s^{-1} when $t = 3$.

    (a) Show that $s = \frac{1}{24} t^2 (42 - t^2)$.

    (b) Find the total distance, in metres, that P moves before returning to O. (ULEAC)

26. Two particles A and B, of masses m and $2m$ respectively, are attached to the ends of a light inextensible string which passes over a smooth fixed pulley. The particles are released from rest with the parts of the string on each side of the pulley hanging vertically. When particle B has moved a distance h it receives an impulse which brings it momentarily to rest. Find, in terms of m, g and h, the magnitude of this impulse. (AEB)

27. A particle of mass 0.2 kg is moving in a straight line under the action of a resistive force which, when the particle has speed v m s^{-1}, is of magnitude kv^3 N, where k is a constant. Given that the force is of magnitude 80 N when $v = 2$, find k. Show that, at time t seconds, v satisfies

$$\frac{\mathrm{d}v}{\mathrm{d}t} = -50v^3.$$

Find the time taken for the speed to decrease from 2 m s^{-1} to 1 m s^{-1}.   (AEB)

28. A cricket ball was thrown by one player and caught at the same height from which it was thrown by another player, 30 m away. The ball moved freely under gravity. The greatest height reached by the ball above the point from which it was thrown was 10 m.

(a) Show that the vertical component of the initial velocity of the ball was 14 m s^{-1}.

Calculate

(b) the time of flight, in s, of the ball,

(c) the speed, in m s^{-1}, with which the ball left the thrower's hand.   (ULEAC)

29. A small sphere R, of mass 0.08 kg, moving with speed 1.5 m s^{-1}, collides directly with another small sphere S, of mass 0.12 kg, moving in the same direction with speed 1 m s^{-1}. Immediately after the collision R and S continue to move in the same direction with speeds U m s^{-1} and V m s^{-1} respectively. Given that $U:V = 21:26$,

(a) show that $V = 1.3$,

(b) find the magnitude of the impulse, in N s, received by R as a result of the collision.   (ULEAC)

30. A vector $\mathbf{v}$ is given by $\mathbf{v} = (t^2 - t)\mathbf{i} + (2t - t^2)\mathbf{j}$, where $\mathbf{i}$ and $\mathbf{j}$ are constant perpendicular unit vectors and t is a variable scalar.
Find expressions, in terms of t, for

(a) $\dfrac{\mathrm{d}\mathbf{v}}{\mathrm{d}t}$,   (b) $\dfrac{\mathrm{d}^2\mathbf{v}}{\mathrm{d}t^2}$,   (c) $\left|\dfrac{\mathrm{d}\mathbf{v}}{\mathrm{d}t}\right|^2$.

   (ULEAC)

31. A particle A has mass 0.5 kg and is acted on by two forces $\mathbf{F}_1$ N and $\mathbf{F}_2$ N. At time $t = 0$ the particle is at rest at the point whose position vector relative to a fixed origin O is $(0.5\mathbf{i})$ m. Given that $\mathbf{F}_1 = 25\mathbf{i} + 20\mathbf{j}$, $\mathbf{F}_2 = 15\mathbf{i} - 20\mathbf{j}$ find the position vector of A at time t seconds.

At time t seconds the position vector of a second particle B relative to O is $(0.75\mathbf{i} + 30t^2\mathbf{j})$ m. Find the position vector of A relative to B at time t seconds and hence, or otherwise, find the time T seconds at which the particles are closest together.

Determine the work done on A by each of the forces $\mathbf{F}_1$ and $\mathbf{F}_2$ during the interval $0 \leqslant t \leqslant T$.   (AEB)

CHAPTER 12

THREE-FORCE EQUILIBRIUM. MOMENT

COPLANAR FORCES IN EQUILIBRIUM

When any number of forces acting on a body are in equilibrium they cause no change of any sort in the motion of the body, i.e.
(a) the resultant force is zero
(b) the set of forces has no turning effect.

In Chapter 6 we considered a *particle* in equilibrium under the action of any number of forces. A particle is regarded as a mass at a *point* so the forces acting on it all pass through that point, i.e. they are concurrent and therefore cannot have any turning effect.
So, as we saw in that chapter, for a particle to be in equilibrium we need only to ensure that the resultant force in each of two directions is zero.

In certain circumstances there are alternative methods for dealing with the equilibrium of a particle and we now take a look at two special cases.

Two Forces in Equilibrium

The resultant of the two forces is zero so they must be of equal magnitude and act in opposite directions.
But if the forces act along parallel lines they have a turning effect.

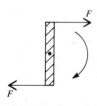

So, for equilibrium, the forces must act in the same straight line.

**Two forces in equilibrium must be equal and opposite
and act in the same straight line.**

THREE FORCES IN EQUILIBRIUM

Consider first the resultant of two non-collinear forces P and Q. We know that when lines representing P and Q are drawn to scale, one after the other, the line joining the starting point to the end point represents the resultant in magnitude and direction; the actual position of the resultant however is through the point of intersection of P and Q.

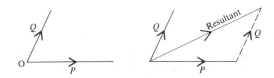

Now if a force R is added to P and Q so that the three forces are in equilibrium, R must cancel out the effect of the resultant of P and Q. Hence R is equal and opposite to this resultant and passes through the point of intersection of P and Q.

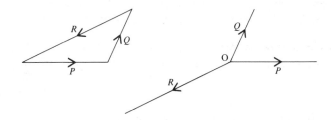

Therefore

**three forces in equilibrium must be concurrent and
can be represented *in magnitude and direction*
by the sides of a triangle taken in order.**

This triangle is known as a *triangle of forces* and it can be used to solve a problem if, in the diagram, there *already is* a triangle whose sides are parallel to the forces acting. Such a triangle is similar to the triangle of forces so the lengths of its sides are proportional to the magnitudes of the corresponding forces.

Do not expect that there always will be a suitable triangle in the diagram. If you cannot spot one, the method used in Chapter 6, of collecting the components in two perpendicular directions and equating to zero in each case, can also be used.

In Chapter 13 we will see how the 'triangle of forces' method can be applied to a rigid body that is in equilibrium under the action of three coplanar forces, (remembering that those forces must be concurrent) but first we will apply it to forces acting on a particle.

Example 12a

A string of length 1 m is fixed at one end to a point A on a wall; the other end is attached to a particle of weight 12 N. The particle is pulled aside by a horizontal force F newtons until it is 0.6 m from the wall. Find the tension in the string and the value of F.

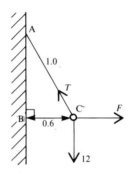

The forces acting on the particle are

In $\triangle ABC$, AB is in the direction of the weight

BC is in the direction of the force

CA is in the direction of the tension

$\therefore$ ABC is similar to the triangle of forces.

$$\therefore \quad \frac{12}{AB} = \frac{F}{BC} = \frac{T}{CA}$$

$\triangle ABC$ is a 3,4,5 triangle, so AB $= 0.8$

$$\therefore \quad \frac{12}{0.8} = \frac{F}{0.6} = \frac{T}{1}$$

$$\Rightarrow \quad F = \frac{7.2}{0.8} = 9 \quad \text{and} \quad T = 15$$

The tension in the string is 15 N and the value of F is 9.

EXERCISE 12a

In each question a particle of weight 10 N is attached to one end A of a light
inextensible string. The other end of the string is attached to a fixed point B.
The particle is held in the given position by the force shown in the diagram.
Copy the diagram and mark the forces acting on the particle. Identify a suitable
triangle of forces and hence find the magnitudes of the force and the tension in
the string. (You may need to extend one or more of the force lines).

1.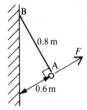

5.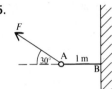

2.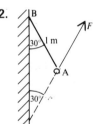

6.

3.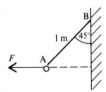

7.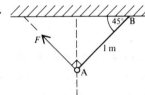

4.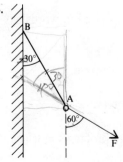

8.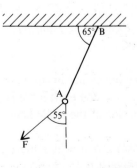

LAMI'S THEOREM

The triangle of forces is useful only when we know some of the lengths involved in the problem. There is an alternative method however, which is applicable when the *angles between the forces* are known rather than any lengths. This method is based on the sine rule.

Consider again three concurrent forces P, Q and R that are in equilibrium, and the corresponding triangle of forces ABC in which BC represents P, CA represents Q and AB represents R.

The angles α, β and γ between the forces are equal to the exterior angles of $\triangle$ABC as shown.

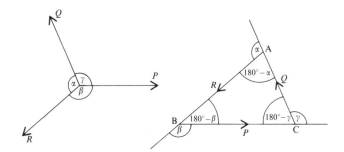

Using the sine rule in $\triangle$ABC gives

$$\frac{P}{\sin(180° - \alpha)} = \frac{Q}{\sin(180° - \beta)} = \frac{R}{\sin(180° - \gamma)}$$

Then, as $\sin(180° - \alpha) = \sin \alpha$, we have

$$\frac{P}{\sin \alpha} = \frac{Q}{\sin \beta} = \frac{R}{\sin \gamma}$$

This relationship is known as *Lami's Theorem* and can be expressed in words as follows:

> **If three forces are in equilibrium they must be concurrent**
> **and each force is proportional to**
> **the sine of the angle between the other two forces.**

Lami's Theorem can give a neat solution to many three-force problems *in which all the angles are known.*

Example 12b

A particle of weight 16 N is attached to one end of a light string whose other end is fixed. The particle is pulled aside by a horizontal force until the string is at 30° to the vertical. Find the magnitudes of the horizontal force and the tension in the string.

Let P newtons and T newtons be the magnitudes of the horizontal force and the tension respectively.

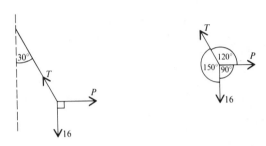

Using Lami's Theorem gives $\quad \dfrac{T}{\sin 90°} = \dfrac{P}{\sin 150°} = \dfrac{16}{\sin 120°}$

Hence $\quad\quad\quad\quad\quad\quad\quad \dfrac{T}{1} = \dfrac{P}{\frac{1}{2}} = \dfrac{16}{\frac{\sqrt{3}}{2}}$

$\Rightarrow \quad\quad\quad T = \dfrac{32}{\sqrt{3}} = \dfrac{32\sqrt{3}}{3} \quad \text{and} \quad P = \dfrac{16}{\sqrt{3}} = \dfrac{16\sqrt{3}}{3}$

Therefore the magnitude of the horizontal force is $\frac{16\sqrt{3}}{3}$ N and the magnitude of the tension is $\frac{32\sqrt{3}}{3}$ N.

A solution to the example above, using the method of resolving in two perpendicular directions, was given on p. 00. We recommend the reader to compare the two methods to see which they prefer.
The two solutions are included to emphasise that there is no right or wrong approach – simply a choice.

EXERCISE 12b

In each question use Lami's Theorem to find the values of P and Q. Give answers corrected to 3 significant figures.

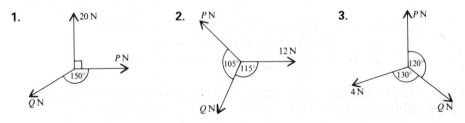

1.

2.

3.

4.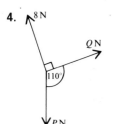

5.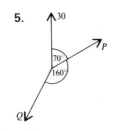

6.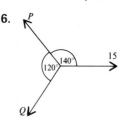

Choice of Method

There are pros and cons for each of the methods we now have for solving three-force equilibrium problems. Here are some points to consider.

- If two of the three forces are perpendicular it is easy to resolve in these two directions and equate the collected components to zero.

- If the given information includes all relevant lengths, and a *suitable triangle is present in the diagram,* using the triangle of forces is quick and easy.

- If all the relevant angles are given, Lami's Theorem at once gives a separate equation for each unknown force.

There is no *best method* for everyone, as individual preference varies, so just remember that *all* the methods work.

EXERCISE 12c

Answer each question by consciously choosing the method *you* think best and then using it.

1.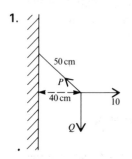

Find *P* and *Q*.

2.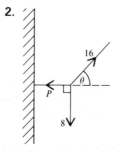

Find *P* and *θ*.

3.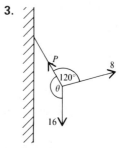

Find *P* and *θ*.

NON-CONCURRENT FORCES

So far we have dealt only with the equilibrium of a particle, i.e. a situation in which the forces are bound to be concurrent; concurrent forces cannot cause rotation so up to now we have not had to worry about turning effect.

When forces act on a rigid body of significant size however, there is no longer any physical reason why these forces should be concurrent. It therefore follows that they are capable of producing rotation and we must now take this into account when considering the equilibrium of a set of forces acting on a rigid body.

First we must take a look at how to measure turning effect.

Consider a rod pivoted at its midpoint P. If it is perfectly uniform, the rod can hang in a horizontal position.

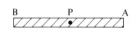

When a downward force *F* is applied at one end A the rod rotates clockwise as shown. The force has not made the rod move bodily downwards, it has caused the rod to *turn about the pivot*.

It can be shown experimentally that an additional force 2*F*, applied downwards halfway along PB, will maintain the rod in its original position. Each force exerts a turning effect on the rod and together they restore the balance of the rod.

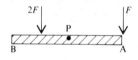

The two forces applied to the rod are not equal however, so clearly the turning effect of a force does not depend entirely on its magnitude.

The other factor is the distance from the pivot of the point of application of the force; a smaller force, further from the pivot can balance a larger force nearer to the pivot.

Experiments show that

the turning effect of a force is given by
magnitude of force × perpendicular distance from pivot

To give a full description of the turning effect of a force we must also give the sense of rotation, i.e. clockwise or anticlockwise.

THE MOMENT OF A FORCE

The turning effect of a force is called the *moment* of the force (or sometimes *torque*).

Not all objects rotate about a pivot, they may turn about a hinge or a fulcrum etc. The general name *axis of rotation* applies to all cases. This name emphasises the fact that rotation does not take place about a point but about a line. The line (axis) is perpendicular to the plane in which the forces act.
So in the examples used above we should really say that the rod rotates about 'a horizontal axis through P'. However it is common practice to refer to *rotation about a point*: it is then taken for granted that the axis of rotation passes through that point and is perpendicular to the plane of the force system.

The Unit of Moment

The magnitude of the moment of a force F, acting at a perpendicular distance d from the axis of rotation, is given by $F \times d$

The unit in which it is measured is the newton metre, $N m$.

It may appear at first sight that this unit could apply in another context, as the work done by a force in moving a particle through a linear distance is also the product of force and distance, suggesting the newton metre as the unit. However, as we always use the joule as the unit of work there is no confusion over the $N m$ which is used exclusively for moments.

The Sign of a Moment

Earlier, when we were collecting components of forces, we chose a *positive direction;* components in that direction had a + sign while components in the opposite direction took a negative sign.

In the same way, we choose a positive sense of rotation when dealing with a system of moments. If, for example, we decide to make anticlockwise the positive sense, an anticlockwise moment has a + sign while a clockwise moment has a negative sign. The resultant moment of a number of forces is then the algebraic sum of the separate moments.

The positive sense does not always have to be anticlockwise; an individual choice can be made for each problem.

Zero Moment

When a force passes through the axis of rotation, its distance from that axis is zero. Therefore the moment of the force about that axis is zero.

Determining the Sense of Rotation

Most people looking at a diagram can see immediately the sense of rotation that a particular force would cause. From experience however we know that there are a few who have a 'blind spot' here. There is a simple ploy for any readers who have this problem:

stick a pin into the point on the diagram about which turning will take place and pull the page (gently!) in the direction of the force. You will then *see* the rotation happening.

Examples 12d

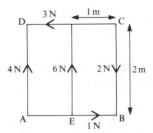

ABCD is a square lamina subjected to the forces shown in the diagram. Find the clockwise moment of each of the forces about an axis through

(a) B (b) A (c) E

(a)	Magnitude of force	1 N	2 N	3 N	4 N	6 N
	⊥ distance from B	0	0	2 m	2 m	1 m
	clockwise moment about B	0	0	−6 N m	8 N m	6 N m

(b)	Magnitude of force	1 N	2 N	3 N	4 N	6 N
	⊥ distance from A	0	2 m	2 m	0	1 m
	clockwise moment about A	0	4 N m	−6 N m	0	−6 N m

(c)	Magnitude of force	1 N	2 N	3 N	4 N	6 N
	⊥ distance from E	0	1 m	2 m	1 m	0
	clockwise moment about E	0	2 N m	−6 N m	4 N m	0

2. The diagram shows a rod AB, free to rotate about the end A. Taking the anticlockwise sense as positive, find the moment about A of each of the forces acting on the rod. Hence find the resultant (total) moment of the forces about A.

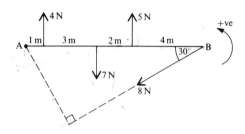

The moment of the 4 N force is $(4 \times 1)\,\text{Nm}$ = $4\,\text{Nm}$
The moment of the 7 N force is $-(7 \times 4)\,\text{Nm}$ = $-28\,\text{Nm}$
The moment of the 5 N force is $(5 \times 6)\,\text{Nm}$ = $30\,\text{Nm}$

The perpendicular distance from A to the force of 8 N is AB sin 30°, i.e. 10 sin 30°.

The moment of the 8 N force is $-(8 \times 10 \sin 30°)\,\text{Nm}$ = $-40\,\text{Nm}$

The resultant moment is $4 + (-28) + 30 + (-40)\,\text{Nm}$ i.e. $-34\,\text{Nm}$

The resultant moment about A is 34 Nm clockwise.

When the collected moments of a number of forces are required about an axis through A, say, we say we are *taking moments about A*. This is denoted by the symbol A↺; the sense of the curved arrow indicates the positive sense of rotation, so A↺ means *taking anticlockwise moments about A*.

3. A force **P**, represented by $4\mathbf{i} + 2\mathbf{j}$, acts through the point with position vector $6\mathbf{i} + \mathbf{j}$, and a second force **Q**, represented by $\mathbf{i} - 3\mathbf{j}$, acts through the point with position vector $2\mathbf{i}$. Given that the units are newtons and metres, find the magnitude and sense of the resultant moment of **P** and **Q** about O.

For each force the components in the **i** and **j** directions are marked on the diagram so that the moment of each component can be found;

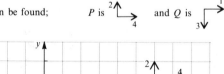

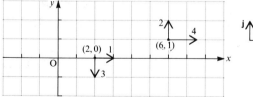

O↺   gives $4 \times 1 - 2 \times 6 + 1 \times 0 + 3 \times 2 = -2$

The resultant moment about O is 2 Nm anticlockwise.

EXERCISE 12d

1. Find the magnitude and sense of the moment of the given force about the point O.

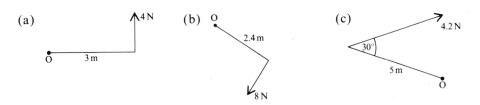

(a)   (b)   (c)

(d)   A force, in newtons, represented by $5\mathbf{i} - 7\mathbf{j}$ and acting through the point with position vector $\mathbf{i} + 3\mathbf{j}$.

2. Find, in magnitude and sense, the resultant moment of the given forces about the point A.

(a)

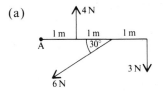

(d)

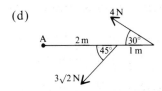

(b)

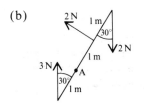

(e)

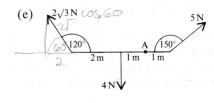

(c)

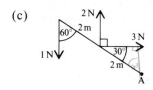

(f)

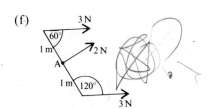

(g)   Forces, in newtons, represented by $2\mathbf{i} - 7\mathbf{j}$ and $-3\mathbf{i} + 4\mathbf{j}$, acting through the points with position vectors $6\mathbf{i} + \mathbf{j}$ and $4\mathbf{j}$ respectively, where $\mathbf{i}$ and $\mathbf{j}$ are unit vectors and A is the origin.

3.

ABCD is a square of side 1 m. Find the magnitude
and sense of the resultant moment of the given forces
about   (a) A   (b) D.

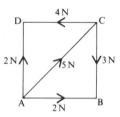

4.

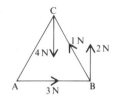

The diagram shows an equilateral triangle of side 2 m.
Find the resultant anticlockwise moment of the forces
shown about

(a) A   (b) C.

5. (a) Find the resultant clockwise moment about a horizontal axis through A, of
the forces acting on the beam AB shown in the diagram.

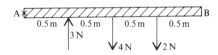

(b) When a force F newtons is applied at B, perpendicular to the beam, the
resultant moment is zero. Find the value of F.

6.

A force represented by $7\mathbf{i} + 4\mathbf{j}$ acts
through the point with position
vector $\mathbf{i} - \mathbf{j}$. The units are newtons and
metres. Find the anticlockwise moment of
the force about an axis through the point
with position vector $2\mathbf{i} + \mathbf{j}$.

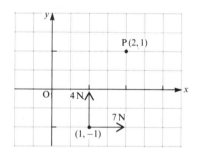

7. Four forces, measured in newtons, are represented by $4\mathbf{i} + 2\mathbf{j}$, $5\mathbf{i}$, $\mathbf{i} - 6\mathbf{j}$,
and $-3\mathbf{j}$. They act respectively through points with position vectors measured
in metres and represented by $\mathbf{i} - \mathbf{j}$, $\mathbf{i} + \mathbf{j}$, $4\mathbf{i}$ and $3\mathbf{j}$.
Find the magnitude and sense of their resultant moment about

(a) the origin O   (b) the point $(1, 1)$.

CHAPTER 13

EQUILIBRIUM OF A RIGID BODY

When a number of concurrent coplanar forces act on a stationary object, the only type of motion they can produce is linear motion (i.e. translation) because since they all act through one point they have no turning effect. If, however, the forces are not concurrent they may also cause the object to rotate.

**The object is in equilibrium if there is no change in the state of its motion.
No change in translation means that the linear resultant of the forces must be zero.
No change in rotation means that the resultant moment of the forces about *any* axis must be zero.**

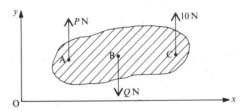

Consider an object that is in equilibrium under the action of the set of parallel forces shown in the diagram.

There are no components of force in the direction Ox, so there can be no change in motion in that direction. Therefore, since the object is in equilibrium, we know that

the resultant force in the direction Oy is zero
the resultant moment about A (or any other point) is zero.

As there are only two unknown forces, we need only two equations to find them. So we can

either   resolve parallel to the forces and take moments about one axis
or   take moments about each of two axes.

Note that, because there is a choice of method, readers can, if they wish, use the method not chosen first as a check.

Note also that if only one unknown force has to be found, one only of the above equations may give sufficient information.

Examples 13a

1. The diagram shows a uniform beam of length 3 m and weight 40 N, suspended in equilibrium in a horizontal position by two vertical ropes, one attached at the end A and the other at C, 1 m from the other end B. Find the tension in each rope. (The weight of a uniform beam acts through the midpoint.)

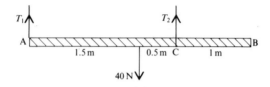

There are no horizontal forces so no information is given by resolving horizontally but as there are only two unknowns we need only two equations. We will choose to take moments about both A and C.

A↺ $\qquad\qquad T_2 \times 2 - 40 \times 1.5 = 0$ $\qquad\qquad$ [1]

C↺ $\qquad\qquad T_1 \times 2 - 40 \times 0.5 = 0$ $\qquad\qquad$ [2]

From [1] $\qquad\qquad\qquad\qquad T_2 = 30$

From [2] $\qquad\qquad\qquad\qquad T_1 = 10$

The tensions in the two ropes are 30 N and 10 N.

Check: Resolving ↑ gives $\quad 30 + 10 - 40 = 0$

2. A uniform plank of weight W and length $6a$ rests on two supports at points B and C as shown in the diagram. The plank carries a load $2W$ at the end A and a load $3W$ at the end D. Find, in terms of W, the force exerted by each support.

B↺ gives $\qquad\qquad 2W \times a + F_2 \times 3a - W \times 2a - 3W \times 5a = 0$ $\qquad\qquad$ [1]

↑ gives $\qquad\qquad F_1 + F_2 - 2W - W - 3W = 0$ $\qquad\qquad$ [2]

From [1] $\qquad\qquad F_2 = 15Wa \div 3a = 5W$

From [2] $\qquad\qquad F_1 = 6W - F_2 = W$

The supporting forces at B and C are W and $5W$ respectively.

Check: C↺ gives

$2W \times 4a - F_1 \times 3a + W \times a - 3W \times 2a = (8 - 3 + 1 - 6)Wa = 0$

3. A scaffold board of weight 50 N and length 4 m lies partly on a flat roof and projects 2 m over the edge. A load of weight 30 N is carried on the overhanging end B and the board is prevented from tipping over the edge by a force applied at the other end A. If the weight of the scaffold board acts through a point 1 m from the end A, what is the value of the least force needed?

The least force will *just* prevent the board from toppling when it is about to lose contact with the roof except at the edge, so the reaction between the board and the roof acts at the edge of the roof.

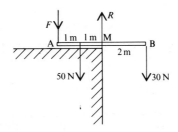

In this problem only the magnitude of the force at A is required and not R. So if we take moments about M, in order to avoid introducing R, only this one equation is needed.

$$M\circlearrowright \text{ gives } \quad 30 \times 2 - 50 \times 1 - F \times 2 = 0$$
$$\Rightarrow \qquad\qquad\qquad\qquad F = 5$$

So the least force needed is 5 N.

Check: $A\circlearrowright$ gives $\quad 50 \times 1 - R \times 2 + 30 \times 4 = 0 \quad \Rightarrow \quad R = 85$

$\uparrow\quad$ gives $\qquad\qquad R - F - 50 - 30 = 0 \quad \Rightarrow \quad F = 5$

EXERCISE 13a

In each question from 1 to 4, a light beam (i.e. the weight is negligible) rests in a horizontal position on two supports, one at A and the other at B, and carries loads as shown. Find the force exerted at each support.

1.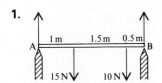

3.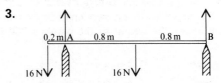

2.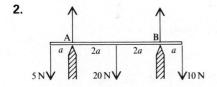

4.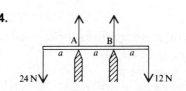

5. A see-saw consists of a light plank of length 4 m, balanced on a fulcrum at the centre. A child of weight 220 N sits on one end.

    (a) How far from the other end should another child, of weight 280 N, sit if the see-saw is to be balanced?

    (b) What force is exerted by the fulcrum?

Questions 6 to 9 concern a horizontal uniform beam supported by vertical ropes. In each case find the tensions in the ropes.

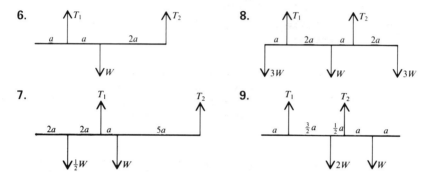

6.

7.

8.

9.

10. A non-uniform plank of wood 3 m long is being carried by two men, one at each end of the plank. Mick is taking a load of 42 N and Tom at the other end is supporting 22 N. Find the distance from Mick's end of the point through which the weight of the plank acts.

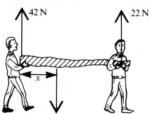

11. A boy builds a simple bridge over a stream by supporting a uniform plank of wood symmetrically on two small brick piers, one on each bank. The piers are 2.6 m apart, the weight of the plank is 300 N and the boy's weight is 420 N.

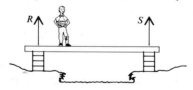

Find the force exerted by each pier when the boy stands

    (a) over one of the piers

    (b) 1 m from one pier

    (c) in the centre of the bridge.

12. A uniform beam AB, of weight 50 N and 2.5 m long, has weights of 20 N and 30 N hanging from the ends A and B respectively. Find the distance from A of the point on the beam where a support should be placed so that the beam will rest horizontally.

13. A uniform plank AB, of mass 80 kg and length 5 m, overhangs a flat roof by 2 m. A boy can walk to within 0.6 m of the overhanging end B when a load of mass 12 kg is placed on the opposite end A.

    (a) What is the mass of the boy?

    (b) Find the smallest extra load that must be placed at the end A to enable the boy to walk right to the end B.

14. A non-uniform rod of weight $3W$ and length $2a$ is suspended by a string attached to the midpoint of the rod. The rod is horizontal when a weight W hangs from one end of the rod. If this weight is removed, find the supporting force that is needed at the opposite end to maintain the rod in its horizontal position.

15. A rod of length 1.2 m is placed on a table top with part of the rod protruding over the edge. The weight of the rod is 10 N. The rod can just hold a particle of weight 6 N, placed on the overhanging end A, without toppling over the edge.

    (a) If the rod is uniform, what length of rod is on the table top?

    (b) If, instead, the weight of the rod acts at a point G, and 0.5 m of the rod protrude over the edge of the table, find the length of AG.

16.

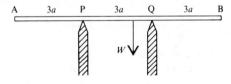

The diagram shows a non-uniform beam AB, of weight W and length $9a$, resting on supports at P and Q. When a load of $\frac{1}{4}W$ is attached at B the beam is just on the point of losing contact with the support at P (i.e. the reaction at P just becomes zero).

    (a) Find the distance from Q of the point through which the weight of the beam acts.

    (b) The load is removed from B and a load Wx is attached at A. If the beam is now just about to lose contact with the support at Q, find the value of x.

17. A light rod of length 2 m is suspended in a horizontal position by a string attached to a point P on the rod. Particles of masses 0.5 kg, 0.8 kg and 1 kg are attached to the rod at points distant 0.4 m, 1 m and 1.2 m from one end A. If $AP = x$ m, find the value of x.

EQUILIBRIUM OF A SET OF NON-CONCURRENT COPLANAR FORCES

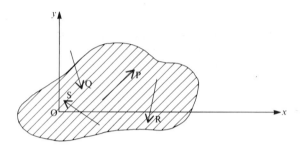

We know that if the forces shown in the diagram act on a body, that body will be in equilibrium only if both the resultant force and the resultant moment are zero.

Now the forces acting on the object are not parallel so the resultant force has components in the directions of both Ox and Oy.

Therefore, if the object is to be in equilibrium the sum of the components in each of the directions Ox and Oy must be zero.

So now we can give the general conditions necessary for an object to be in equilibrium under the action of a set of non-parallel coplanar forces, i.e.

the resultant force in the direction Ox is zero

the resultant force in the direction Oy is zero

the resultant moment about any axis is zero.

Applying these conditions to a particular problem gives three equations, so three unknown quantities can be found.

In some problems it is more convenient to use an alternative set of three independent equations, i.e.

the resultant in the direction Ox (or Oy but not both) is zero

the resultant moment about a particular axis is zero

the resultant moment about a different axis is also zero.

The following Examples and Exercise show how the *general* method can be applied to a variety of questions.

Examples 13b

1.

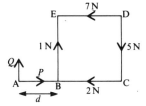

The diagram shows a set of forces in equilibrium. BCDE is a square of side 0.5 m. Calculate P, Q and d.

First we will resolve parallel to BC and CD

$\rightarrow \qquad P - 2 - 7 = 0 \qquad \Rightarrow \qquad P = 9$

$\uparrow \qquad Q + 1 - 5 = 0 \qquad \Rightarrow \qquad Q = 4$

Now taking moments about B gives a simple equation for d.

$\text{B} \circlearrowleft \qquad Q \times d + 5 \times 0.5 - 7 \times 0.5 = 0$

$\Rightarrow \qquad\qquad\qquad\qquad 4d - 1 = 0 \qquad \Rightarrow \qquad d = 0.25$

2. A ladder of length 4 m and weight W newtons rests in equilibrium with its foot A on horizontal ground and resting against a vertical wall at the top B. The ladder is uniform; contact with the wall is smooth but contact with the ground is rough and the coefficient of friction is $\frac{1}{3}$. Find the angle θ between the ladder and the wall when the ladder is on the point of slipping.

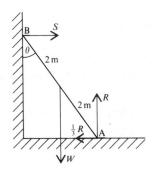

The ladder is about to slip so friction is limiting and $F = \mu R$.

Resolving horizontally and vertically gives

$\rightarrow \qquad S - \frac{1}{3}R = 0$

$\uparrow \qquad R - W = 0$

Hence $R = W$ and $S = \frac{1}{3}W$

The third necessary equation is given by taking moments about *any* axis. The best choice of axis is through A because both F and R pass through A and therefore have zero moment.

A) $\quad S \times 4 \cos \theta - W \times 2 \sin \theta = 0$

Hence $\qquad S = \frac{1}{2} W \tan \theta$

$\therefore \qquad \frac{1}{3} W = \frac{1}{2} W \tan \theta$

$\Rightarrow \qquad \tan \theta = \frac{2}{3}$

The angle between the ladder and the wall is $34°$ (nearest degree).

3. A uniform beam, of length $8a$ and weight W, is hinged to a fixed point at one end A and bears a load of $2W$ at the other end B. The beam is held in a horizontal position by a rope of length $5a$ joining the midpoint G of the beam to a point C vertically above A. Find the tension in the rope and the magnitude of the force exerted by the hinge.

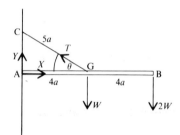

In $\triangle CAG$, AG $= 4a$ and CG $= 5a$

$$\cos \theta = \frac{AG}{CG} = \frac{4a}{5a} = \frac{4}{5}$$

i.e. $\qquad \Rightarrow \qquad \sin \theta = \frac{3}{5}$

The direction of a force at the hinge is not known so we can use components X and Y. If we take moments about A, both components of the hinge force have zero moment so T can be found directly.

A) $\quad W \times a + 2W \times 2a - T \times a \sin \theta = 0$

$\Rightarrow \qquad\qquad T = 5W \div \sin \theta = \dfrac{25W}{3}$

Now we resolve horizontally and vertically.

$\rightarrow \qquad\qquad X - T \cos \theta = 0 \qquad \Rightarrow \qquad X = \dfrac{4T}{5} = \dfrac{20W}{3}$

$\uparrow \qquad Y + T \sin \theta - W - 2W = 0 \qquad \Rightarrow \qquad Y = 3W - \dfrac{3T}{5} = = -2W$

The magnitude of the hinge force is given by $\sqrt{(X^2 + Y^2)}$

i.e. $\qquad \sqrt{\left(\dfrac{400W^2}{9} + 4W^2 \right)} = \dfrac{2W\sqrt{109}}{3}$

The magnitude of the hinge force is $\frac{2}{3} W \sqrt{109}$

The tension in the string is $\dfrac{25W}{3}$

EXERCISE 13b

1. In questions 1 to 6 a uniform ladder of weight 200 N and length 2 m rests with one end A on level ground and the other end B resting against a vertical wall. When the ladder is in limiting equilibrium (i.e. is just about to slip) the angle between the ladder and the wall is θ. Where appropriate give answers corrected to 3 significant figures and angles to the nearest degree.

2.

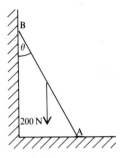

3.

Contact with the wall is smooth; contact with the ground is rough and the coefficient of friction is $\frac{1}{3}$. Find θ.

Contact with the wall is smooth; contact with the ground is rough and the coefficient of friction is μ. If $\theta = 45°$ find

4.
  (a)   the normal reaction with the ground

  (b)   the frictional force

  (c)   the value of μ.

Contact is rough both with the wall and with the ground; the coefficient of friction in both cases is $\frac{1}{4}$. Find θ. (Remember that the ladder will not slip until friction is limiting at both points of contact.)

Contact with the wall is smooth and contact with the ground is rough. When $\theta = 60°$ a workman of weight 80 kg can climb one quarter of the way up the ladder before limiting equilibrium is reached. Find the reaction at the wall and the coefficient of friction. Take g as 10 and treat the workman as a point-load.

If, in question 4, θ is reduced to 30°, find how far up the ladder the man can now climb.

If, in question 4, contact is rough at both ends of the ladder and $\mu = \frac{1}{3}$ in each case, find θ if the workman can just climb to the top of the ladder. Take g as 10.

In questions 7 to 11 a uniform rod PQ of length 2 m and weight 24 N is hinged at the end P to a fixed point and is in equilibrium. R is a point vertically above P.

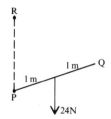

7. PQ is kept horizontal by a support at Q. Find the magnitude of the supporting force and the magnitude and direction of the force exerted on the rod by the hinge.

8. PQ is held at an angle of 60° to the upward vertical through P, by a light string joining Q to R. Given that QR = 2 m, find

(a) the tension in the string

(b) the magnitude of the reaction at the hinge.

9. PQ is held at an angle of 60° to the downward vertical through P by a horizontal force F newtons applied at the end Q. Find the value of F.

10. PQ is held at an angle of 60° to the upward vertical at P by a light string of length 1 m joining R to the midpoint of the rod. Find

(a) the tension in the string

(b) the magnitude and direction of the force exerted on the rod at the hinge.

11. PQ is horizontal. A light string of length 2.5 m connects Q to R and a load of 20 N is applied to the rod at the end Q. Find

(a) the tension in the string

(b) the magnitude and direction of the force acting on the rod at the hinge.

12. A uniform rod AB of length $4a$ rests with the end A in rough contact with level ground where the coefficient of friction is $\frac{1}{2}$. A point C on the rod, distant $3a$ from A, rests against a smooth peg. The rod is in limiting equilibrium when it is at 30° to the ground. Find, in terms of W

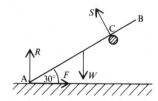

(a) the reaction at the peg

(b) the frictional force.

13.

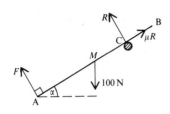

A uniform rod rests in rough contact with a peg at C and a force F newtons acts at A as shown in the diagram. The weight of the rod is 100 N, $\tan \alpha = \frac{3}{4}$, BC = CM = a and MA = 2a. Find

(a)   the coefficient of friction at the peg

(b)   the value of F.

14. A uniform plank of weight 54 N is hinged to a fixed point at one end A. The plank has to be held at 30° to the downward vertical by a force applied at the end B. Find the magnitude of the force needed if it is applied

(a)   at right angles to the plank

(b)   horizontally.

USING 'THREE-FORCE EQUILIBRIUM'

The equilibrium problems in the previous exercise were solved by the general method which uses resolving and taking moments.

There are, however, alternative methods for solving problems involving the equilibrium of a rigid body under the action of three – or sometimes more than three – coplanar forces.

We know that if three forces are in equilibrium they must be concurrent. So if a rigid body is in equilibrium under the action of three forces, and the point of intersection of two of the forces is known, the third force *must* pass through the same point. In this way a force whose direction is not normally known (e.g. the force at a hinge) can be located on a diagram.

This principle can be extended to cover some problems where more than three forces are involved initially but can be reduced in number.

At a point of rough contact, for example, when friction is limiting, the resultant reaction can be used instead of normal reaction plus frictional force, so reducing the number of forces acting.

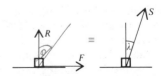

Once it is established that three forces are keeping a body in equilibrium, the triangle of forces or Lami's theorem can be applied, providing some elegant solutions as well as variety of method.

Examples 13c

The first example shows how the resultant reaction at a point of rough contact can be used to reduce a four-force problem to a three-force case allowing the angle of friction to be found.

1. A uniform ladder AB, resting with B in rough contact with the ground and A in smooth contact with a wall, is just about to slip when inclined at 30° to the wall. Find the coefficient of friction at the ground.

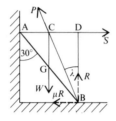

If we use the resultant reaction at B the three forces then acting on the ladder are concurrent at C. CG and DB are parallel and G bisects AB (the rod is uniform), so C bisects AD.

As P is the resultant of R and μR, angle CBD is the angle of friction.

$$\tan \lambda = \frac{CD}{DB} = \frac{\frac{1}{2}AD}{DB} = \frac{1}{2}\tan 30°$$

$$\therefore \qquad \tan \lambda = \frac{1}{2} \times \frac{1}{\sqrt{3}} = \frac{\sqrt{3}}{6} = \mu$$

The coefficient of friction is $\frac{\sqrt{3}}{6}$.

The second example shows how the direction of a force that is not generally known can be located in a three-force problem and mensuration used to calculate this direction.

2. A uniform rod is hinged at one end A to a wall. The other end B is pulled aside by a horizontal force until the rod is in equilibrium at 60° to the wall. Find the direction of the hinge force.

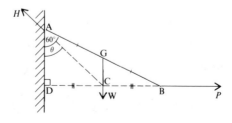

The weight of a *uniform* rod acts through the midpoint of the rod. The horizontal force and the weight of the rod meet at the point C. Three forces in equilibrium must be concurrent, so the hinge force also acts through C, i.e. the direction of H is along CA.

The angle θ can be found from the mensuration of the general diagram. GC is parallel to AD and G bisects AB so C bisects DB (Intercept Theorem).

In $\triangle$ADC $\tan \theta = \dfrac{DC}{AD} = \dfrac{\frac{1}{2}DB}{AD}$

$$= \tfrac{1}{2} \tan 60° = \tfrac{1}{2}\sqrt{3}$$

$\therefore$ $\quad\quad\quad\quad\quad\quad \theta = 40.9°$ (3 sf)

The hinge force is inclined at 40.9° to the wall.

The next example again uses the concurrency of three forces in equilibrium and applies Lami's Theorem since all angles are known.

3. One end A of a uniform rod AB of weight 36 N is hinged to a fixed point. The rod is held in a horizontal position by a string connecting B to a point C vertically above the hinge. If the string is inclined at 45° to the rod, find

(a) the direction of the force exerted on the rod by the hinge

(b) the tension in the string.

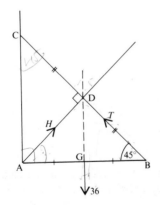
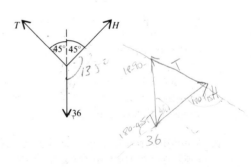

(a) The weight and the tension meet at D. G bisects AB so D bisects BC.

$\therefore$ $\quad\quad\quad$ A$\hat{\text{D}}$B = 90° $\quad\quad \Rightarrow \quad\quad$ D$\hat{\text{A}}$B = 45°

The direction of the hinge force is at 45° to the rod.

(b) Using Lami's Theorem,

$$\frac{T}{\sin 135°} = \frac{36}{\sin 90°} \quad\left(= \frac{H}{\sin 135°}, \text{ not wanted}\right)$$

$\Rightarrow \quad\quad T = 36 \times 0.7071 = 25.5$

The tension is 25.5 N (3 sf)

The questions in the next exercise can each be solved in a variety of ways. The following points may help you to choose the method you wish to use.

If there are only three forces acting, always make sure first that they are concurrent on the diagram. This should be done regardless of the method.

- If what is required is an angle, it may be that it can be found from the diagram without any further work.

- If there is an obvious triangle of forces and its sides are known or easy to find, it is suitable to use the equal ratios of force to length of side.

- If only three forces act and the angles between them are known, Lami's Theorem immediately gives the magnitude of each force separately.

- If there are more than three forces, or you do not care for any of the special methods, the general approach of resolving and taking moments can be applied to any problem. It may sometimes take a little longer but it always works!

EXERCISE 13c

In questions 1 and 2 a uniform rod AB of weight W, is hinged to a fixed point at A and held in the position shown by a force F.

(a) Draw a diagram showing the direction of the force exerted on the rod by the hinge.

(b) Find, in terms of W, the magnitude of the hinge force.

1.

2.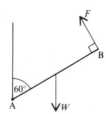

3. The end A of a uniform rod AB is in rough contact with a horizontal surface. The end B rests on the rim of a smooth disc fixed in the same vertical plane as the rod, so that AB is a tangent to the disc and is inclined at 30° to the horizontal. The angle of friction at A is λ and the rod is just about to slip.

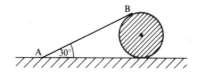

(a)   Draw a general diagram and on it mark *three* forces that keep the rod in limiting equilibrium and the angle of friction.

(b)   *Write down* Lami's Theorem for the three forces.

4. A uniform rod AB has one end B in rough contact with the ground. The rod rests against a smooth rail C and is about to slip. If $\tan \alpha = \frac{1}{2}$ and $\mu = \frac{1}{2}$, show that the resultant reaction at B is equal to the reaction at C.

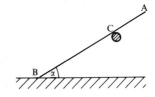

For questions 5 to 8 use this diagram, in which AB is a uniform ladder of weight W and is just about to slip.

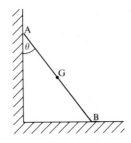

5. Contact at A is smooth, contact at B is rough and $\mu = \frac{3}{4}$. Find θ and the reaction at A.

6. Both contacts are rough and $\theta = 45°$. Find μ. (The ladder will not slip until limiting friction is reached at both ends.)

7. Both contacts are rough with $\mu = \frac{1}{2}$. Find θ.

8. Explain why it is impossible for a ladder to rest in equilibrium at any angle to a wall if the wall and the ground are both smooth.

9. The diagram shows a uniform rod AB of weight W, held at right-angles to a wall by a string BC which is inclined at 30° to the wall. Contact with the wall is rough and friction is limiting.

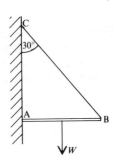

(a)   Mark on a diagram the forces acting on the rod, using the resultant reaction at A.

(b)   Find the angle of friction λ and hence the value of μ, the coefficient of friction between the rod and the wall. (Remember that $\mu = \tan \lambda$)

(c)   Find, in terms of W, the tension in the string.

10. The diagram shows a uniform sphere of weight 20 N, with radius 3 cm, and centre O. The sphere rests in contact with a point A on a smooth wall and is held in position by a string of length 2 cm attached to a point B on the sphere and a point C on the wall.

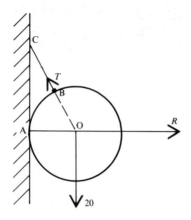

(a) Explain why the direction of the string CB passes through O.

(b) Write down the lengths of AO and CO and find the length of CA.

(c) Identify a suitable triangle of forces and use it to find the tension in the string and the reaction between the sphere and the wall.

CHAPTER 14

CENTRE OF GRAVITY AND CENTRE OF MASS

CENTRE OF GRAVITY

When we consider a rod of weight W, it is obvious that the rod is made up of a large number of very short lengths of material, each with its own weight.

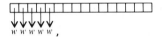

However, when dealing with an equilibrium problem involving a rod we usually mark a singe weight acting at a particular point on the rod. (In Chapter 13 we frequently referred to 'the point through which the weight of the rod acts'.)

Now if we want to replace all the components of weight by a single weight we must ensure that it has exactly the same effect on the rod as the separate components have, i.e.

the total weight is the sum of all the component weights

and

the single weight acts through a point such that the moment of the single weight about any axis is equal to the resultant moment of the components.

The point through which the resultant weight of a body acts is called the centre of gravity of the body and is very often denoted by G and its coordinates by $(\bar{x}, \bar{y})$

The position of the centre of gravity of any object can be found by equating the resultant moment of all the parts, to the moment about the same axis of the total weight acting through the centre of gravity.

This principle applies to any object of any size, shape or dimension but we will look at a simple example of a set of separate particles.

Consider three particles, A, B and C, of weights 4 N, 2 N and 3 N respectively, attached to points on a light rod PQ as shown in the diagram below.

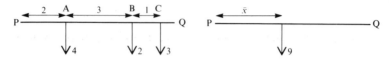

We know that, about any axis, the total moment of the weights of the particles is equal to the moment of the total weight.

Choosing to take moments about P we have

$$4 \times 2 + 2 \times 5 + 3 \times 6 = (4+2+3) \times \bar{x}$$

Hence $\bar{x} = 4$

This approach can be extended to any number of weights, W_1, W_2, W_3, ..., at distances x_1, x_2, x_3, ... from the chosen axis.

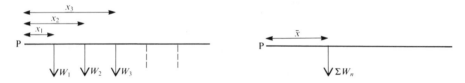

P) $W_1 \times x_1 + W_2 \times x_2 + W_3 \times x_3 + \ldots = \bar{x} \times (W_1 + W_2 + W_3 + \ldots)$

i.e. $\Sigma W_n x_n = \bar{x} \Sigma W_n$

(Σ means 'the sum of terms of this form when n takes values 1, 2, 3 ...')

CENTRE OF MASS

Using $W = mg$ in the equation $\Sigma W_n x_n = \bar{x} \Sigma W_n$ gives

$$\Sigma m_n g x_n = \bar{x} \Sigma m_n g$$

If we take the value of g as constant and cancel it from each term in the moment equation above, we get $\Sigma m_n x_n = \bar{x} \Sigma m_n$ in which each term is of the form mass × its distance from a particular axis.

The solution of this equation is the location of the point G which we have so far called the centre of gravity, i.e. the point about which the *weight* of an object is evenly distributed.

However, as we now see, the *mass* also is evenly distributed about G which therefore can also be called the *centre of mass*.

This argument depends on the assumption that the value of g is the same for all the masses (and so can be cancelled). However, provided that we are not dealing with bodies (e.g. a mountain) that are so high that the value of g changes between the highest and lowest parts, we can make this assumption, i.e.

for an object of normal size, the centre of mass coincides with the centre of gravity

Hence, to find the centre of mass of a particular object we form an equation in which, on one side we have the sum of terms like $m_1 \times x_1$, and on the other side the sum of all the masses multiplied by $\bar{x}$. This equation can be written

$$\Sigma\, m_n \times x_n \;=\; \bar{x} \times \Sigma\, m_n$$

Examples 14a

1. The diagram shows a set of three particles of masses 5 kg, 2 kg and 4 kg attached to a light rod at the given positions. Find the distance from O of the centre of mass, G, of the particles.

Using $\qquad\qquad \Sigma\, m_n \times x_n \;=\; \bar{x} \times \Sigma\, m_n$ gives

$$5 \times 1 + 2 \times 5 + 4 \times 10 \;=\; \bar{x} \times (5 + 2 + 4)$$

$\Rightarrow \qquad\qquad\qquad\qquad\qquad \bar{x} = 5$

The distance of G from O is 5 m.

2. Particles of masses $3m$, $2m$, $6m$, and am are attached to a light rod AB at distances $3d$, $4d$, $8d$, and $12d$ from A respectively. If G, the centre of mass of the particles, is distant $7d$ from A find the value of a.

Using $\qquad \Sigma\, m_n \times x_n \;=\; \bar{x} \times \Sigma\, m_n$ gives

$$3m \times 3d + 2m \times 4d + 6m \times 8d + am \times 12d \;=\; (11+a)m \times 7d$$

$\therefore \qquad\qquad\qquad\qquad (65 + 12a)md \;=\; (11+a)m \times 7d$

$\Rightarrow \qquad\qquad\qquad\qquad\qquad (12-7)a \;=\; 77 - 65 \;=\; 12$

$\Rightarrow \qquad\qquad\qquad\qquad\qquad\qquad a \;=\; 2.4$

EXERCISE 14a

1. Find the distance from P of the centre of mass, G, of two particles placed at P and Q, where PQ is of length 24 cm, if the masses of the particles are respectively

(a) 1 kg and 2 kg   (c) 3 kg and 8 kg   (e) m and $3m$

(b) 3 kg and 1 kg   (d) 6 kg and 4 kg   (f) $8m$ and $6m$.

In each question from 2 to 10 find the distance from A of the centre of mass of the given set of particles.

2.
```
       3a      a      3a       a
A————————o——o—————————o——B
        2m  5m        2m
```

5.
```
         2a           3a        a
A————————————o——————————————o————B
            4m                 m
```

3.
```
     a    a    a      2a
A——o——o——o————————o——B
    2m   m   2m      m
```

6.
```
        a           2a
A o——————————o————————————————o B
    m          m               2m
```

4.
```
         3a         a    a
B o——————————————o——o————— A
   m            2m   2m
```

7.
```
          4a          a
B o————————————————o————o A
   6m             4m    m
```

8.
```
   o 3m
4a |
   o 6m
2a |
   o 7m
3a |
   o 4m
   A
```

9.
```
   A
   |a
   o 2m
   |a
   o 3m
   |a
   o 4m
   |a
   o 5m
   |a
   o 6m
```

10.
```
   o 2m
5a |
   o 8m
4a |
   |A
   a
```

11. Three particles whose masses are 2 kg, 5 kg and x kg are placed at points with coordinates $(1, 0)$, $(2, 0)$ and $(6, 0)$ respectively.

(a) If G, the centre of mass of the particles, is at the point $(3, 0)$, find x.

(b) A fourth particle of mass 2 kg is placed at the point $(5, 0)$. Find the centre of mass of the set of four particles.

12. A light rod AB is 80 cm long. Three particles, each with mass 2 kg, are attached at points distant 20 cm, 36 cm and 48 cm from the end A.

(a) Find the distance from A to G, the centre of mass of the particles.

(b) A fourth particle is attached at B. Find its mass if the centre of mass of the four particles is at the midpoint of the rod.

(c) If the particle distant 48 cm from A is then removed, find the distance of the centre of mass of the remaining three particles from A.

THE CENTRE OF MASS OF PARTICLES IN A PLANE

In all the problems above, the particles concerned lay in a straight line. Now we consider particles situated anywhere in a plane. For convenience we will locate the positions of the particles by coordinates in an xy plane.

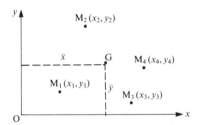

In order to locate G, both its x and y coordinates are required.

The x-coordinate, $\bar{x}$, is given by $\Sigma m_n \times x_n = \bar{x} \times \Sigma m_n$

The y-coordinate, $\bar{y}$, is given by $\Sigma m_n \times y_n = \bar{y} \times \Sigma m_n$

Examples 14b

1. Particles A, B and C of masses 4 kg, 7 kg and 5 kg are placed respectively at points with coordinates $(4, 2)$, $(0, 6)$ and $(1, 5)$. Find the centre of mass of the particles.

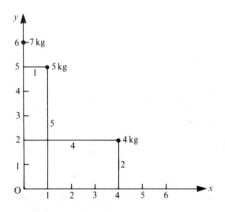

Using $\Sigma m_n \times x_n = \bar{x} \times \Sigma m_n$ gives

$$4 \times 4 + 7 \times 0 + 5 \times 1 = \bar{x}(4 + 7 + 5) \qquad \Rightarrow \qquad \bar{x} = 21 \div 16$$

Using $\Sigma m_n \times y_n = \bar{y} \times \Sigma m_n$ gives

$$4 \times 2 + 7 \times 6 + 5 \times 5 = 16\bar{y} \qquad \Rightarrow \qquad \bar{y} = 75 \div 16$$

The centre of mass is at the point $(\frac{21}{16}, \frac{75}{16})$

2. Particles of masses 4 kg, 2 kg, 5 kg and 6 kg are placed respectively at the vertices A, B, C and D of a light square lamina of side 2 m. Find the distances of the centre of mass of the particles from AB and AD.

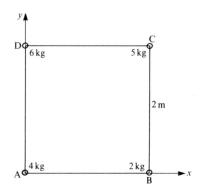

Using $\Sigma m_n \times x_n = \bar{x} \times \Sigma m_n$

$$4 \times 0 + 2 \times 2 + 5 \times 2 + 6 \times 0 = \bar{x} \times (4 + 2 + 5 + 6)$$

$\Rightarrow$ $\bar{x} = 14 \div 17$

Using $\Sigma m_n \times y_n = \bar{y} \times \Sigma m_n$

$$4 \times 0 + 2 \times 0 + 5 \times 2 + 6 \times 2 = \bar{y} \times (4 + 2 + 5 + 6)$$

$\Rightarrow$ $\bar{y} = 22 \div 17$

The distances of the centre of mass are $\frac{14}{17}$ m from AD and $\frac{22}{17}$ m from AB.

Note that from AB (a *horizontal* line) the distance measured is *vertical,* i.e. $\bar{y}$. (In giving the answer to the question above it is dangerously easy to reverse the lines from which the distances are given.)

3. Five particles whose masses are $2m$, $3m$, $2m$, $4m$ and m, are placed at points with position vectors $3\mathbf{i} + \mathbf{j}$, $\mathbf{i} - 4\mathbf{j}$, $5\mathbf{i} + 6\mathbf{j}$, $-\mathbf{i} + 2\mathbf{j}$, and $3\mathbf{i}$ respectively. Find the position vector of their centre of mass.

Let the position vector of the centre of mass be $a\mathbf{i} + b\mathbf{j}$.

Using $\Sigma m_n \times x_n = \bar{x} \times \Sigma m_n$

$$2m \times 3 + 3m \times 1 + 2m \times 5 + 4m \times (-1) + m \times 3 = 12m \times a$$

$\Rightarrow$ $a = 18 \div 12 = \frac{3}{2}$

Using $\Sigma m_n \times y_n = \bar{y} \times \Sigma m_n$

$$2m \times 1 + 3m \times (-4) + 2m \times 6 + 4m \times 2 + m \times 0 = 12m \times b$$

$\Rightarrow$ $b = 10 \div 12 = \frac{5}{6}$

The position vector of the centre of mass is $\frac{3}{2}\mathbf{i} + \frac{5}{6}\mathbf{j}$.

Note that, when the particles are located by their position vectors, the separate x and y equations can be combined in the form

$$\Sigma m_n \times \mathbf{r}_n = \bar{\mathbf{r}} \times \Sigma m_n$$

where $\bar{\mathbf{r}} = \bar{x}\mathbf{i} + \bar{y}\mathbf{j}$

The solution can then be presented more concisely as

$$2m(3\mathbf{i}+\mathbf{j}) + 3m(\mathbf{i}-4\mathbf{j}) + 2m(5\mathbf{i}+6\mathbf{j}) + 4m(-\mathbf{i}+\mathbf{j}) + m(3\mathbf{i}) = 12m\bar{\mathbf{r}}$$

$\Rightarrow$ $\quad\quad (18\mathbf{i}+10\mathbf{j})m = 12m\bar{\mathbf{r}}$

$\Rightarrow$ $\quad\quad\quad\quad \bar{\mathbf{r}} = \frac{3}{2}\mathbf{i} + \frac{5}{6}\mathbf{j}$

EXERCISE 14b

In each question find the coordinates of the centre of mass of the given set of particles.

1.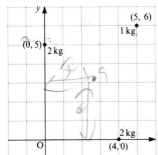

3.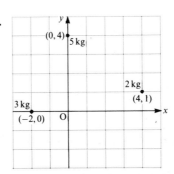

2.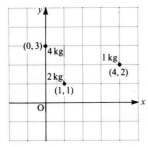

4.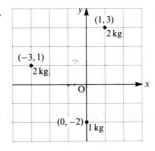

5. The coordinates of the vertices of a light square framework are $(4, 1)$, $(9, 5)$, $(3, 8)$ and $(0, 4)$. Particles of masses 2 g, 4 g, 2 g and 2 g respectively are placed at the vertices.

6. A light L-shaped wire ABC is in the xy plane with particles of equal mass m attached to A, B and C. The coordinates of A, B and C are $(-3, 2)$, $(2, 2)$ and $(2, -1)$ respectively.

THE CENTRE OF MASS OF A UNIFORM LAMINA

The number of particles that make up a lamina is infinitely large, so clearly it is not practical to locate its centre of mass in the same way as for a small number of distinct particles. However there are alternative methods for dealing with some uniform laminas (the mass per unit area of a uniform lamina is constant throughout). One of these methods uses the properties of symmetry.

If a uniform lamina has a symmetrical shape, the mass is equally distributed about the line of symmetry. Therefore the centre of mass must lie somewhere on the line of symmetry. If there is more than one axis of symmetry, it follows that G is located at the point of intersection of these axes.

The Centre of Mass of a Uniform Rod

The midpoint of a uniform rod is clearly its centre of mass, as the masses of the two halves are equal and equally distributed.

The Centre of Mass of a Uniform Square Lamina

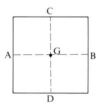

From symmetry, the centre of mass of the square lies somewhere on the line AB that bisects the square because the distribution of the mass is the same on both sides of this line. Similarly the centre of mass lies on CD and therefore is at the midpoint of each of these lines, which is the geometric centre of the square.

The Centre of Mass of a Uniform Rectangular Lamina

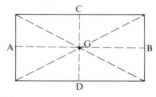

Using symmetry as we did above, the centre of mass is at the point of intersection of AB and CD, i.e. the point where the bisectors of the sides meet.
(Note that this is also the point of intersection of the diagonals of the rectangle.)

The Centre of Mass of a Uniform Triangular Lamina

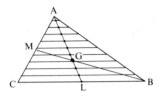

As there is no line of symmetry, we will divide the triangle into strips parallel to one side, BC say, and regard each strip as a 'rod'.

The centre of mass of each strip is at its midpoint, so the centre of mass of the triangle, G, lies on the line passing through all these midpoints, i.e. on the median AL.

Now using strips parallel to AC shows that G also lies on the median BM.

So G is at the point of intersection of the medians of the triangle.

It is a geometric property of a triangle that its medians intersect at a point which is $\frac{1}{3}$ of the way from base to vertex on any median. This point is called the *centroid* of a triangle so we can say that the centre of mass of a uniform triangular lamina is at the centroid of the triangle.

Hence for a right-angled triangle, G is $\frac{1}{3}$ of the way from the right-angle along each of the perpendicular sides.

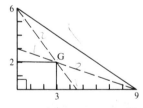

If the lamina is not uniform its centre of mass is unlikely to be at the centroid. In other words, the centroid of a triangle is *always* $\frac{1}{3}$ of the way up any median, but the centre of mass of a triangular lamina is *not necessarily* at the centroid.

Readers may occasionally find centroid used as though it *meant* centre of mass. This is not so. The centroid is a geometric point and *does not change* if the density of the shape varies, whereas the centre of mass is a property of balance and *does* depend on the distribution of the mass.

THE CENTRE OF MASS OF A COMPOUND LAMINA

Now that the position of the centres of mass of a number of common laminas are known, they can be used to find the centre of mass of a uniform lamina that is made up of these shapes.

For each part of the lamina, the mass can be expressed as the product of area and mass per unit area, i.e. mass = area × density.

It follows that the mass of each part, and hence the mass of the whole, is a multiple of ρ, where ρ, pronounced ro, is the symbol for density.

Examples 14c

1. Find the centre of mass of the uniform lamina OABCDEF.

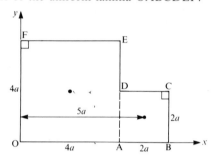

Let ρ be the mass per unit area of the lamina and let $G(\bar{x}, \bar{y})$ be the centre of mass of the whole lamina.

The mass of OAEF is $16a^2\rho$ and its centre of mass is at $(2a, 2a)$

The mass of ABCD is $4a^2\rho$ and its centre of mass is at $(5a, a)$

The mass of OABCDEF is $16a^2\rho + 4a^2\rho = 20a^2\rho$

Using $\Sigma m_n x_n = \bar{x}\Sigma m_n$ gives

$$16a^2\rho \times 2a + 4a^2\rho \times 5a = 20a^2\rho \times \bar{x}$$

Now $4a^2\rho$ cancels giving $\qquad \bar{x} = \frac{13a}{5}$

Similarly using $\Sigma m_n y_n = \bar{y}\Sigma m_n$ we have

$$16a^2\rho \times 2a + 4a^2\rho \times a = 20a^2\rho \times \bar{y} \qquad \Rightarrow \qquad \bar{y} = \frac{9a}{5}$$

The centre of mass of the whole lamina is the point $(\frac{13a}{5}, \frac{9a}{5})$.

The solution of this example can be set down more concisely by using a table as shown below.

Portion	Mass	Coords of G		mx	my
		x	y		
+ OAEF	$16a^2\rho$	$2a$	$2a$	$16a^2\rho \times 2a$	$16a^2\rho \times 2a$
+ ABCD	$4a^2\rho$	$5a$	a	$4a^2\rho \times 5a$	$4a^2\rho \times a$
OABCDEF	$20a^2\rho$	$\bar{x}$	$\bar{y}$	$20a^2\rho \times \bar{x}$	$20a^2\rho \times \bar{y}$

The plus signs are a reminder that the two parts are *added* to give the whole.

Working down the 5th column $16a^2\rho \times 2a + 4a^2\rho \times 5a = 20a^2\rho \times \bar{x}$

$\Rightarrow$ $\bar{x} = \frac{13a}{5}$

Working down the 6th column $16a^2\rho \times 2a + 4a^2\rho \times a = 20a^2\rho \times \bar{y}$

$\Rightarrow$ $\bar{y} = \frac{9a}{5}$

The tabular layout is recommended for all problems of this type.

2. Find the centre of mass of the uniform lamina ABCDE shown in the diagram.

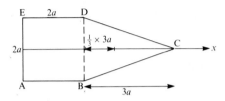

The shape is symmetrical about the line through C that bisects AE so G lies on this line. We will make this line the *x*-axis. Only the *x*-coordinate of the centre of mass is unknown.

Portion	Mass	x-coord. of G	$m_n x_n$
+ ABDE	$4a^2\rho$	a	$4a^2\rho \times a$
+ BCD	$3a^2\rho$	$2a + a$	$3a^2\rho \times 3a$
ABCDE	$7a^2\rho$	$\bar{x}$	$7a^2\rho \times \bar{x}$

$\Sigma\, m_n x_n = \bar{x}\, \Sigma\, m_n$, i.e. working down the last column gives

$4a^2\rho \times a + 3a^2\rho \times 3a = 7a^2\rho \times \bar{x}$

$\Rightarrow$ $\bar{x} = \frac{13a}{7}$

The centre of mass of the lamina is on the line that bisects AE and DB and distant $\frac{13a}{7}$ from AE.

3.

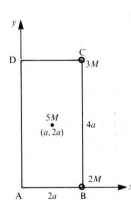

The diagram shows a uniform rectangular lamina ABCD of mass $5M$. A particle of mass $2M$ is attached to B and a particle of mass $3M$ to C. Find the distances from AB and AD of the centre of mass of the lamina complete with its loads.

		Coords of G			
Portion	Mass	x	y	mx	my
+ ABCD	$5M$	a	$2a$	$5Ma$	$10Ma$
+ Particle B	$2M$	$2a$	0	$4Ma$	0
+ Particle C	$3M$	$2a$	$4a$	$6Ma$	$12Ma$
Loaded lamina	$10M$	$\bar{x}$	$\bar{y}$	$10M\bar{x}$	$10M\bar{y}$

Using $\ \Sigma m_n x_n \ = \ \bar{x} \, \Sigma m_n \ $ gives

$$5Ma + 4Ma + 6Ma \ = \ 10M\bar{x} \qquad \Rightarrow \qquad \bar{x} \ = \ 1.5a$$

Using $\ \Sigma m_n y_n \ = \ \bar{y} \, \Sigma m_n \ $ gives

$$. \ \ 10Ma + 0 + 12Ma \ = \ 10M\bar{y} \qquad \Rightarrow \qquad \bar{y} \ = \ 2.2a$$

The centre of mass is distant $1.5a$ from AD and $2.2a$ from AB.

4.

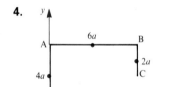

A uniform wire is bent into the shape of three sides of a trapezium, as shown in the diagram. Find the coordinates of the centre of mass of the shape.

The centre of mass of each section of the wire is at the midpoint and the mass of the wire is proportional to its length. We will use k as a constant of proportion.

		Coords of G			
Portion	Mass	x	y	mx	my
+ OA	$4a \times k$	0	$2a$	$4ka \times 0$	$4ka \times 2a$
+ AB	$6a \times k$	$3a$	$4a$	$6ka \times 3a$	$6ka \times 4a$
+ BC	$2a \times k$	$6a$	$3a$	$2ka \times 6a$	$2ka \times 3a$
OABC	$12ak$	$\bar{x}$	$\bar{y}$	$12ak \times \bar{x}$	$12ak \times \bar{y}$

Using $\ \Sigma mx \ = \ \bar{x} \, \Sigma m, \ $ where $\ \Sigma m \ = \ (4+6+2)ak, \ $ gives

$$4ka \times 0 \ + \ 6ka \times 3a \ + \ 2ka \times 6a \ = \ 12ka\bar{x}$$

$$\Rightarrow \qquad\qquad\qquad\qquad\qquad \bar{x} \ = \ \tfrac{5a}{2}$$

Using $\ \Sigma my \ = \ \bar{y} \, \Sigma m \ $ gives

$$4ka \times 2a \ + \ 6ka \times 4a \ + \ 2ka \times 3a \ = \ 12ka\bar{y}$$

$$\Rightarrow \qquad\qquad\qquad\qquad\qquad \bar{y} \ = \ \tfrac{19a}{6}$$

The coordinates of the centre of mass are $\left(\tfrac{5a}{2}, \tfrac{19a}{6}\right)$.

5.

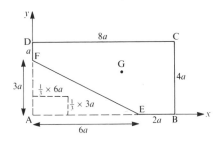

A triangle AEF is cut off from the rectangular lamina ABCD as shown in the diagram. Find the position of G, the centre of mass of the remaining lamina.

The centre of mass for the triangular part is $\frac{1}{3}$ of $6a$ and $\frac{1}{3}$ of $3a$ from A along AB and AD.

Portion	Mass	Coords of G		mx	my
		x	y		
+ ABCD	$32a^2\rho$	$4a$	$2a$	$32a^2\rho \times 4a$	$32a^2\rho \times 2a$
− AEF	$9a^2\rho$	$2a$	a	$9a^2\rho \times 2a$	$9a^2\rho \times a$
EBCDF	$23a^2\rho$	$\overline{x}$	$\overline{y}$	$23a^2\rho \times \overline{x}$	$23a^2\rho \times \overline{y}$

This time the triangle is *taken away* from the rectangle so we subtract the mass of the triangle from that of the rectangle. This is indicated by the negative sign in the second row.

$\Sigma m_n x_n = \overline{x} \Sigma m_n$ gives

$$32a^2\rho \times 4a - 9a^2\rho \times 2a = 23a^2\rho \times \overline{x} \quad \Rightarrow \quad \overline{x} = \frac{110a}{23}$$

$\Sigma m_n y_n = \overline{y} \Sigma m_n$ gives

$$32a^2\rho \times 2a - 9a^2\rho \times a = 23a^2\rho \times \overline{y} \quad \Rightarrow \quad \overline{y} = \frac{55a}{23}$$

The distances of G from AD and AB are respectively $\frac{110a}{23}$ and $\frac{55a}{23}$.

Note that whenever the position of G has been found it should be marked on the diagram, to see whether it looks about right.

EXERCISE 14c

In this exercise all laminas and rods are uniform.

1. State the coordinates of the centre of mass of each lamina.

(a)

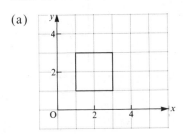

(b)

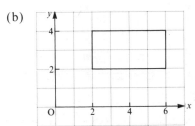

(c)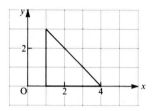

(d)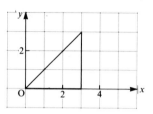

2. Write down the coordinates of the centre of gravity of each section of the given shape.

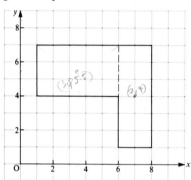

(a)

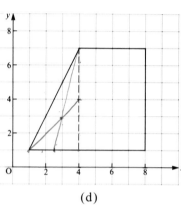

(d)

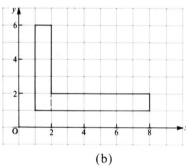

(b)

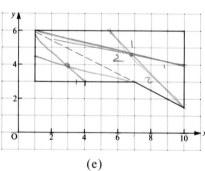

(e)

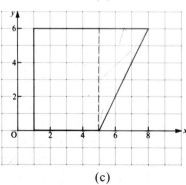

(c)

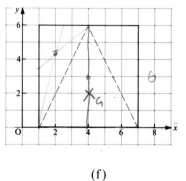

(f)

Keep your solutions to questions 3 to 8 as you will need them in Exercise 14d.

In questions 3 to 6, choose axes based on symmetry and find the coordinates of the centre of gravity of each lamina. Take the side of one square as 1 unit.

3.

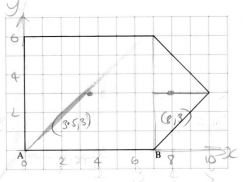

5.

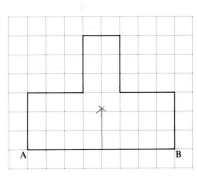

4.

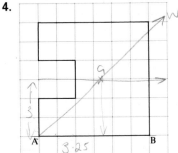

6.

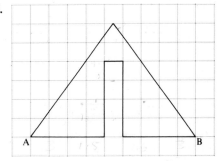

Find the centre of gravity of each framework of rods. All rods are of equal density.

7.

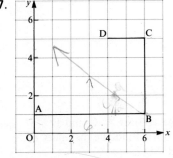

8.

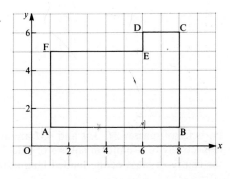

Find the centre of gravity of each lamina.

9.

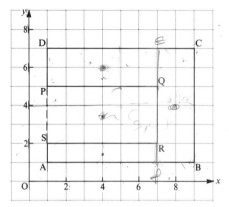

11.

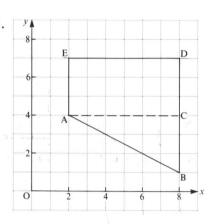

10.

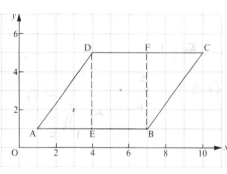

12.

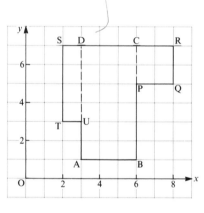

13. The diagram shows a uniform right-angled triangular lamina of mass $3M$, loaded at B with a particle of mass $2M$.
Find the coordinates of the centre of mass of the triangle complete with its load.

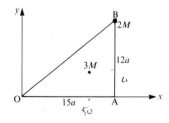

14. Particles of masses M, $3M$ and $4M$ are attached to the vertices A, C and D respectively of a uniform square of side $2a$ and mass $6M$. Find the distances of the centre of mass of the loaded lamina from AB and AD.

SUSPENDED LAMINAS

When an object is suspended by a string attached to one point of the object, two forces act on the object; the tension in the string vertically upwards and the weight of the object vertically downwards.
If the object hangs at rest it is in equilibrium and the two forces must therefore be in the same line, i.e. the centre of gravity, and therefore the centre of mass, is vertically below the point of suspension.

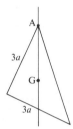

Now if the object is one whose centre of mass is known, a right-angled triangular lamina for example, the position in which the lamina will hang can be found by joining the point of suspension, A say, to G.
In equilibrium, AG is vertical.

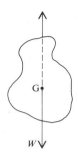

Suppose that we are asked to find the angle between the vertical and one of the sides of the lamina.
It is not always easy to see how to calculate this angle from the diagram of the lamina in its suspended position. It is often more straightforward, if it is possible, to draw the lamina so that two of its sides are horizontal and vertical, and on this diagram mark the line which would be vertical.

As an example consider the right-angled triangle shown above.

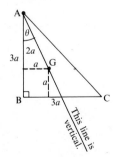

We know that AG is vertical and that G is distant a from both BA and BC.
Therefore to find the angle θ between AB and the vertical we use

$$\tan \theta = \frac{a}{2a} = \frac{1}{2}$$

$$\Rightarrow \qquad \theta = 27° \text{ (nearest degree)}$$

In this example, because the centre of mass of a triangular lamina is quotable, the position of G is known at the outset.

Now consider the compound lamina given in the first worked problem in Examples 14c.

Clearly the suspended position cannot be dealt with until G is located, i.e. all the work done in the example has to be carried out before we can begin to consider how it hangs in equilibrium when suspended freely from a specified point, O say.

Now we have already drawn a diagram to locate G and, *on the same diagram,* we can draw OG, i.e.

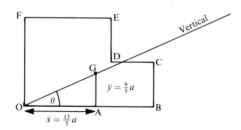

It is immediately clear that $\tan \theta = \dfrac{\bar{y}}{\bar{x}} = \dfrac{9a}{5} \div \dfrac{13a}{5} = \dfrac{9}{13}$

Although there may be some readers who prefer to work with the 'suspended position' diagram, the majority will find it easier to use the approach given above.

Example 14d

A uniform lamina ABCD is in the shape of a trapezium in which $AB = 4a$, $AD = 3a$, $DC = a$ and angle DAB is $90°$.

(a) Find the distance of G, the centre of mass of the lamina, from AB and AD.

(b) The lamina is suspended from A and hangs freely. What is the angle between AB and the vertical?

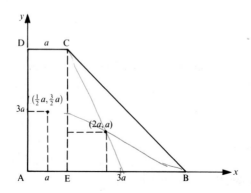

(a)

Portion	Mass	Coords of Centre of Mass		mx	my
		x	y		
+ AECD	$3a^2\rho$	$\frac{1}{2}a$	$\frac{3}{2}a$	$3a^2\rho \times \frac{1}{2}a$	$3a^2\rho \times \frac{3}{2}a$
+ EBC	$\frac{9}{2}a^2\rho$	$2a$	a	$\frac{9}{2}a^2\rho \times 2a$	$\frac{9}{2}a^2\rho \times a$
ABCD	$\frac{15}{2}a^2\rho$	$\bar{x}$	$\bar{y}$	$\frac{15}{2}a^2\rho \times \bar{x}$	$\frac{15}{2}a^2\rho \times \bar{y}$

Using $\Sigma m_n x_n = \bar{x}\,\Sigma m_n$

$$3a^2\rho \times \frac{1}{2}a + \frac{9}{2}a^2\rho \times 2a = \frac{15}{2}a^2\rho \times \bar{x}$$

$\Rightarrow \qquad\qquad \bar{x} = \dfrac{7a}{5} = $ distance of G from AD

Using $\Sigma m_n y_n = \bar{y}\,\Sigma m_n$

$$3a^2\rho \times \frac{3}{2}a + \frac{9}{2}a^2\rho \times a = \frac{15}{2}a^2\rho \times \bar{y}$$

$\Rightarrow \qquad\qquad \bar{y} = \dfrac{6a}{5} = $ distance of G from AB

(b) When ABCD is hanging freely from A, AG is vertical, so the angle we want is $\hat{GAB}$.

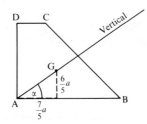

$$\tan \alpha = \frac{\bar{y}}{\bar{x}} = \frac{6a}{5} \div \frac{7a}{5} = \frac{6}{7}$$

$\Rightarrow \qquad \alpha = 41°$ (nearest degree)

EXERCISE 14d

Refer to Exercise 14c for questions 1 to 3.

When freely suspended from A, find the angle between AB and the vertical for the lamina given in

1. Exercise 14c, question 3.

2. Exercise 14c, question 4.

3. Exercise 14c, question 5.

4. When the framework of rods given in Exercise 14c question 7 is suspended freely from B, find the angle between AB and the vertical.

5. The framework of rods given in Exercise 14c question 8 is suspended freely from B. Find the angle that BC makes with the vertical.

6. A framework of rods is in the shape of a rectangle ABCD in which AB = 6a and BC = 4a. For each of the rods AB, BC and AD, the mass of a length a of the rod is m; the rod DC is light. A particle of mass 8m is attached to C and a particle of mass m is attached to B.

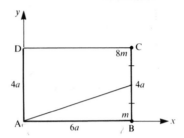

(a) Find the distances from AB and AD of the centre of mass of the loaded framework.

(b) Show that when the framework is suspended freely from A, the line joining A to the midpoint of BC is vertical.

7. When the lamina shown in the diagram is freely suspended from A, AB hangs at an angle α to the vertical where $\tan \alpha = \frac{10}{9}$.

(a) Find, in terms of a and d the coordinates of the centre of mass of the lamina.

(b) Express d in terms of a.

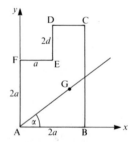

Note that any reader who would like further practice in this topic can use any of the questions in Exercise 14c.

STABILITY OF A LAMINA ON AN INCLINED PLANE

Consider a lamina in the shape of a trapezium ABCD, whose centre of gravity is G, placed in a vertical plane with the side AB on an inclined plane, as shown. (We must assume that the lamina can be balanced in this position). For the lamina to be in equilibrium, contact with the plane must be rough, therefore three forces act on the lamina and these are:

its weight W, the normal contact force R which is perpendicular to the plane and a frictional force F acting up the plane.

If the lamina is in equilibrium these three forces must be concurrent, so R and F must act at the point P where the line of action of the weight meets the plane.

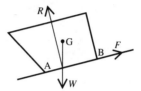

Now let us consider three different positions as the angle of inclination of the plane is gradually increased.

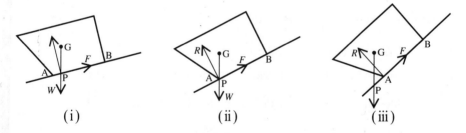

(i)   (ii)   (iii)

In diagram (i), the point P is somewhere between A and B; the normal reaction can act through that same point and the lamina can be in equilibrium.

In diagram (ii), P coincides with A, i.e. P is at the end-point of the base; the normal reaction can act through A and the lamina is *just* in equilibrium. This diagram therefore shows the critical angle of inclination of the plane.

When the inclination of the plane is further increased, to a position as shown in diagram (iii), P does not lie within the base. The normal reaction can only act through a point on AB so it is impossible in this case for W and R and F to be collinear. Instead the weight exerts a turning effect about A, i.e. the lamina is *not* in equilibrium and will topple about A.

MODULE E – *Mechanics 1* contains printing errors on pages 297 and 298 – below is the correct version of page 298.

298   **Chapter 14**

Examples 14e

ABC is a uniform triangular lamina in which   AB $= 0.3$ m, AC $= 0.6$ m and $\hat{A} = 90°$. The lamina is balanced in a vertical plane with the side AB in contact with a rough plane inclined at an angle θ to the horizontal. Given that the lamina can rest in equilibrium with A below B find,

(a) the greatest possible value of θ

(b) the least possible value of the coefficient of friction, μ.

(a) The greatest possible value of θ is reached when the weight passes through A.

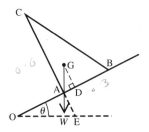

The distances of G from A are:

  0.1 m parallel to AB,   0.2 m perpendicular to AB.

Therefore $\tan \text{AGD} = \dfrac{0.1}{0.2}$ $\Rightarrow$ $\widehat{\text{AGD}} = 26.5\ldots°$

Triangles OAE and GAD are similar so $\widehat{\text{AGD}} = \theta$

Therefore, to a whole number of degrees, the greatest safe value of θ is $26°$.

(b) Consider the equilibrium of the lamina for a general value of θ.

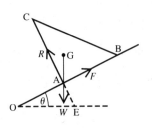

Resolving

$\nearrow$ $F - W \sin \theta = 0$

$\nwarrow$ $R - W \cos \theta = 0$

$\therefore$ $\dfrac{F}{R} = \dfrac{W \sin \theta}{W \cos \theta} = \tan \theta$

But $\dfrac{F}{R} \leqslant \mu$ $\Rightarrow$ $\tan \theta \leqslant \mu$

When θ has its greatest possible value, $\tan \theta = \frac{1}{2}$ therefore $\mu \geqslant \frac{1}{2}$
Therefore the least value of μ is 0.5

Consider a lamina in the shape of a trapezium ABCD, whose centre of gravity is G, placed in a vertical plane with the side AB on an inclined plane, as shown. (We must assume that the lamina can be balanced in this position). For the lamina to be in equilibrium, contact with the plane must be rough, therefore three forces act on the lamina and these are:

its weight W, the normal contact force R which is perpendicular to the plane and a frictional force F acting up the plane.

If the lamina is in equilibrium these three forces must be concurrent, so R and F must act at the point P where the line of action of the weight meets the plane.

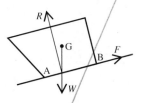

Now let us consider three different positions as the angle of inclination of the plane is gradually increased.

raster = "rgp297-2b"\raster = "rgp297-2c"

(i)   (ii)   (iii)

In diagram (i), the point P is somewhere between A and B; the normal reaction can act through that same point and the lamina can be in equilibrium.

In diagram (ii), P coincides with A, i.e. P is at the end-point of the base; the normal reaction can act through A and the lamina is *just* in equilibrium. This diagram therefore shows the critical angle of inclination of the plane.

When the inclination of the plane is further increased, to a position as shown in diagram (iii), P does not lie within the base. The normal reaction can only act through a point on AB so it is impossible in this case for W and R and F to be collinear. Instead the weight exerts a turning effect about A, i.e. the lamina is *not* in equilibrium and will topple about A.

Examples 14e

ABC is a uniform triangular lamina in which   AB $= 0.3$ m, AC $= 0.6$ m and
$\widehat{A} = 90°$. The lamina is balanced in a vertical plane with the side AB in contact
with a rough plane inclined at an angle θ to the horizontal. Given that the
lamina can rest in equilibrium with A below B find,

(a)   the greatest possible value of θ

(b)   the least possible value of the coefficient of friction, μ.

(a)   The greatest possible value of θ is reached when the weight passes through A.

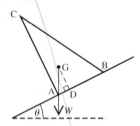

The distances of G from A are:

0.1 m parallel to AB,   0.2 m perpendicular to AB.

Therefore tan AGD $= \dfrac{0.1}{0.2}$ $\Rightarrow$ $A\widehat{G}D = 26.5...°$

Triangles OAE and GAD are similar so $A\widehat{G}D = \theta$

Therefore, to a whole number of degrees, the greatest safe value of θ is 26°.

(b)   Consider the equilibrium of the lamina for a general value of θ.

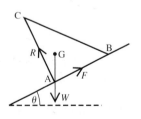

Resolving

$\nearrow$ $-W\ \hbox{sin}\ \theta\, = \,0$

$\nwarrow$ $-W\ \hbox{cos}\ \theta\, = \,0$

$\therefore$ $\dfrac{F}{R} = \dfrac{W \sin \theta}{W \cos \theta} = \tan \theta$

But $\dfrac{F}{R} \leqslant \mu$ $\Rightarrow$ $\tan \theta \leqslant \mu$

When θ has its greatest possible value, $\tan \theta = \frac{1}{2}$ therefore $\mu \leqslant \frac{1}{2}$
Therefore the least value of μ is 0.5

A uniform square lamina of side 2a is placed in a vertical plane with one side
on a plane inclined at an angle α to the horizontal.
Find the largest possible value of α for which the lamina can remain in
equilibrium, assuming that the plane is rough enough to prevent sliding.

A uniform rectangular lamina ABCD is such that $AB = 0.5\,m$ and $BC = 0.3\,m$.
The lamina is placed with one of its sides on a rough inclined plane.
The plane of the lamina is vertical.
Find the maximum angle of inclination of the plane to the horizontal, and the
least coefficient of friction between the lamina and the plane, for which the
lamina can rest in equilibrium without toppling or sliding if the side in contact
with the plane is

(a)   AB   (b) BC.

The uniform lamina ABC is placed, as shown
in the diagram, on an inclined plane rough
enough to prevent the lamina from sliding down.
When the angle of inclination, α, of the plane
to the horizontal is such that $\tan \alpha = \frac{3}{4}$,
the lamina is on the point of toppling about A.
Find the value of h.

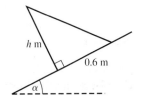

ABC is a uniform lamina in the shape of an isosceles triangle in which
$AB = AC = 0.6\,m$ and the length of BC is $2a$ metres. The lamina is placed, in a
vertical plane, with BC in contact with a rough inclined plane for which the
coefficient of friction is $\frac{1}{2}$. If the lamina is on the point of sliding and toppling
simultaneously find

(a)   the angle at which the plane is inclined to the horizontal

(b)   the value of a.

CHAPTER 15

MODELLING REAL LIFE

USING APPROPRIATE MODELS

Throughout this book we have introduced 'real life' situations such as ladders leaning against walls, cars travelling along roads and so on. The mathematical techniques we used for calculating unknown quantities were based on known physical laws. The situation was simplified, however, by ignoring the effects of certain quantities in the interest of simplicity (this is often referred to as 'making assumptions'). This process is called mathematical modelling.

The results obtained from such a model can only be approximate. There are several reasons for this: firstly assumptions such as treating a string as 'light', i.e. weightless, have an effect on the accuracy of results, also the laws of mechanics that we use are not true in an absolute sense but they give results that are accurate enough for most purposes. Secondly we are dealing with physical quantities such as time, distance, weight etc., and these cannot be measured exactly.

In this section we ask for any assumptions to be stated and for an assessment of the effect these have on the accuracy of results.

Suppose that you want to find the speed at which you can throw a ball. Speed cannot be measured directly but when you throw the ball both the time for which the ball is in the air and the distance it travels before landing can be measured. This information can then be used to calculate a value for the initial speed of the ball if some assumptions are made so that an appropriate model can be used.

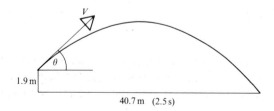

If the ball is thrown and the measurements taken are as shown in the diagram, then the assumptions we need to make are:

there is no air resistance,

there is no wind to affect speed,

the ball is small enough to be treated as a particle.

the measurements taken are accurate to as many figures as are given.

This situation can be modelled by the methods for a particle projected in still air.

We also need to take a value for g. The acceleration due to gravity depends on where it is measured. It varies between $9.83\,\mathrm{m\,s}^{-2}$ at the poles and $9.78\,\mathrm{m\,s}^{-2}$ at the equator but, to 2 significant figures, we can use $g = 9.8$ anywhere on the surface of the earth. The value of g also depends on the position relative to the surface of the earth but, unless the location is more than about 10 km above or below sea-level, the value remains 9.8 correct to 2 significant figures.

So, using the equations of motion for a projectile as our model, when $t = 2.5$, $x = 40.7$ and $y = -1.9$ we have

$$40.7 = (V \cos \theta) \times 2.5$$

and $\quad -1.9 = (V \sin \theta) \times 2.5 - 0.5 \times 9.8 \times (2.5)^2$

from which we can calculate V as $19.92\ldots$

Now we need to consider the accuracy of the result; this will depend on how reasonable the assumptions are and on the accuracy of the measurements. These are some of the factors involved.

- The amount of air resistance depends on the ball; a cricket ball is likely to be subject to less than is a new fluffy tennis ball so the results using a cricket ball will be more accurate.

- Wind speed can affect motion (it is taken into consideration when confirming athletics records). The stronger the wind is, the less accurate are the results.

- Unless you are standing on a championship quality bowling green, or equivalent, when you throw the ball, the ground is unlikely to be perfectly horizontal. On the other hand, obvious slopes can be seen and avoided, so this assumption will have little effect on the result.

- A large ball such as a beachball cannot reasonably be treated as a particle.

- The measurements taken depend on the instruments used. Using a tape to measure the distance, a measurement to the nearest 10 cm is reasonable. For the time (using an ordinary stop watch and the help of another person) it is probable that a measurement to the nearest tenth of a second can be made.

Only the person performing this experiment can judge these criteria, but assuming that there is no wind, that the ground is level, that the ball is smooth and not large and that measurements are taken carefully, then it is reasonable to assume that we have used an appropriate model and that the speed is about $20\,\mathrm{m\,s^{-2}}$.

EXERCISE 15a

This exercise contains some suggestions for practical work.

1. Look at Examples 2b, number 1, on page 40. List as many as you can of the simplifications and assumptions that have been made in the solution of this problem.

2. Repeat question 1 with reference to Examples 5d, number 2.

3. Now consider Examples 7a, number 2. One simplification has been made that is not stated in the question. Reword the question so that this simplification is either implied or stated.

4. Read Examples 8c, number 3; some simplifications are stated in the question. Are there others that are assumed in the solution? Is it reasonable to say that, when A reaches the edge of the table, the speed of each block is between $2.25\,\mathrm{m\,s^{-1}}$ and $2.35\,\mathrm{m\,s^{-1}}$?

5. A hoist consists of a rope passing over a pulley with a loaded platform tied to one end. The other end of the rope is attached to a winch that winds or unwinds the rope to raise or lower the platform. This is modelled by considering a light string, passing over a smooth pulley, with a particle on one end representing the loaded platform and a fixed thin rod on the other end representing the winch. The model is used to calculate the tension in the rope when different loads are raised with an acceleration of $0.5\,\mathrm{m\,s^{-2}}$. List three assumptions that need to be made for this model to be appropriate.

6. Suppose that the model used for the worked example refered to in question 4 was used in this situation.
A puck, with a teflon coated surface to reduce friction to a minimum, is placed on a horizontal table. A string attached to the puck passes over the edge of the table and weights are hung on the free end of the string.

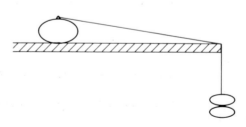

Give two further assumptions that need to be made if the model is to be appropriate.

Comment on the accuracy of the answer obtained from the model when applied to this situation.

7. Try the experiment described above this exercise for yourself (you will need some help with the timing). Try it first with a golf ball (or equivalent) and then with a tennis ball. Comment on how your assumptions affect your results.

8. You are given a steel disc and a sheet of vinyl floor covering. You are asked to find the coefficient of friction between the disc and the sheet. Describe how this could be done. List any assumptions that you make.

9. A model car is powered by a small electric motor. How could you find the maximum power of the engine and how accurate would your answer be? If you have access to such a car, try the experiment and describe a way in which you could check your result.

 10.

The diagram shows someone climbing a ladder. The weight of the ladder, the person's weight and the coefficients of friction at the contacts at the ends of the ladder are all known. Describe the model that you would use to work out the maximum inclination of the ladder to the vertical for it to be safe for the person to climb to the top of the ladder. State any simplifications that you would need to make and whether the resulting answer would err on the side of safety or otherwise.

11. A device for testing the strength of windscreen glass in aircraft, fires 'birds' horizontally at the screen. The 'birds' have a mass of about 4 kg. Describe a model that could be used to work out a value for the impulse on the screen when the 'bird' hits it. State all simplifications and assumptions that are needed for your model to be appropriate.

12. At least five people, each with a stop watch, are needed to collect the data for this experiment.

Position people at measured intervals (about 5 m apart is reasonable) from a set of traffic lights. Record the time taken for a car to reach each person from a stationary start at the lights.

Use the data to find how the acceleration of the car varied over the chosen distance.

Estimate how accurate your results are and suggest ways in which the accuracy of the results could be improved.

13. A bullet is fired at point blank range into a piece of wood. Describe how the deceleration of the bullet might be found.

14. The specification of a motor bike quotes a mass of 200 kg, maximum power rating of 45 kW and a maximum speed of 190 km/h. Janet had a test ride on this bike and found that the maximum speed which she could achieve on a level racing track was 170 km h^{-1}. Comment on possible reasons for this difference.

15. Consider this problem:
A car whose mass is 500 kg, has a power rating of 30 kW and a maximum speed of 120 km h^{-1} on the level. Find the maximum speed of the car down an incline of 20%.

(a) What assumptions would you need to make to get a model from which you could find a value for the maximum speed.

(b) Explain why any answer, using just the information given, would be unrealistic.

16. An aeroplane, flying at a steady 250 km h^{-1}, at a height of 50 m, is to drop aid parcels which, it is hoped, will land in a designated target area. How far horizontally from the target area should the drop be made? Give two reasons why it is not possible to be certain that the parcels will land in the target area.

CONSOLIDATION D

SUMMARY

Equilibrium of three Coplanar forces

- The forces must be concurrent.
- The resultant force must be zero.
- The forces can be represented in magnitude and direction by the sides of a triangle taken in order.
- Lami's Theorem applies, i.e. each force is proportional to the sine of the angle between the other two forces.

Moment

The moment of a force F about an axis is $F \times d$ where d is the perpendicular distance from the axis to the line of action of the force.

General Equilibrium of Coplanar forces

The resultant force must be zero and the resultant moment about any axis must also be zero.

For a rigid body in equilibrium under the action of a set of forces, the collected components of the forces in any direction total zero, and the resultant moment about any axis is zero.

Centre of Gravity

The centre of gravity is the point (usually denoted by G) through which the resultant weight acts.

For a suspended object, G is vertically below the point of suspension.

Centre of Mass

The distance from an axis of the centre of mass of a set of masses m_1, m_2, m_3, etc, placed at distances d_1, d_2, d_3, etc from that axis, is defined as $\dfrac{\Sigma m_n \times d_n}{\Sigma m_n}$

MISCELLANEOUS EXERCISE D

In this exercise use $g = 9.8$ unless another instruction is given.

Each question from 1 to 3 is followed by several suggested responses. Choose the correct response.

1. Three forces F_1, F_2 and F_3 are in equilibrium, therefore,

A $\mathbf{F}_1 = \mathbf{F}_2 + \mathbf{F}_3$

B $\mathbf{F}_1 + \mathbf{F}_2 + \mathbf{F}_3 = 0$

C $\mathbf{F}_1 - \mathbf{F}_2 - \mathbf{F}_3 = 0$

2. A light rod, pivoted at A, has forces applied to it as indicated. The rod will:

A   rotate clockwise

B   rotate anticlockwise

C   remain horizontal.

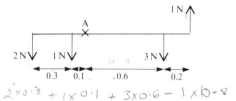

$2 \times 0.3 + 1 \times 0.1 + 3 \times 0.6 - 1 \times 0.8$

3. Three particles of masses 1 kg, 2 kg, 1 kg are at the points whose position vectors are $\mathbf{i} + \mathbf{j}$, $2\mathbf{i} - \mathbf{j}$, $3\mathbf{i} + \mathbf{j}$. The position vector of their centre of mass is:

A $\frac{1}{4}(6\mathbf{i} + \mathbf{j})$ B $2\mathbf{i}$ C $2\mathbf{i} + \mathbf{j}$.

4. A footbridge across a stream consists of a uniform plank AB, of length 5 m and mass 140 kg, supported at the ends A and B. A man of mass 100 kg is standing at a point C on the footbridge. Given that the magnitude of the force exerted by the support at A is twice the magnitude of the force exerted by the support at B, calculate

(a) the magnitude, in N to 3 significant figures, of the force exerted by the support at B

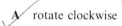

(b) the distance AC.   (ULEAC)

5. ABCD is a rectangle with sides AB and CD 3 m long and sides BC and AD 4 m long. Forces of magnitude 11 N, 3 N, 15 N and 7 N act along BA, BC, CA and CD respectively, the sense of the force being that indicated by the order of the letters. Find the moment of this system of forces about C and the components parallel to BA and BC of the single resultant force to which this system can be reduced.   (WJEC)

6. A uniform rod AB of weight 40 N and length 1.2 m rests horizontally in equilibrium on two smooth pegs P and Q, where   AP = 0.2 m   and BQ = 0.4 m. Find the reactions at P and Q. Find also the magnitude of the greatest vertical load that can be applied at B without disturbing the equilibrium.   (AEB)

7. A force **P** in the direction $4\mathbf{i} + 3\mathbf{j}$ and of magnitude 10 N acts through the point with position vector $(2\mathbf{j})$ m relative to an origin O. Another force **Q** in the direction **j** and of magnitude 2 N acts through the point with position vector $(4\mathbf{i})$ m relative to O. Find

(a) **P** and **Q** in the form $\alpha\mathbf{i} + \beta\mathbf{j}$ where α, β are scalars to be determined in each case

(b) the magnitude of the resultant of **P** and **Q**

(c) the magnitude of the total moment of **P** and **Q** about O. State whether the moment is clockwise or anticlockwise.   (AEB)

8. A non-uniform ladder of length 6 m is supported at its ends on the shoulders of two men, Alf and Ben. Given that the ladder is horizontal and the forces on the shoulders of Alf and Ben are of magnitude 50 N and 70 N respectively, calculate the distance, in metres, of the centre of mass of the ladder from Alf.
                                                                                          (ULEAC)

9. Forces of 5 N, 20 N act along the sides AB and BC respectively of a square of side 1 m. Find the forces acting along the sides AD, DC and the direction CA which together with the given forces will produce equilibrium.   (WJEC)

10. Three particles of masses 0.1 kg, 0.2 kg and 0.3 kg are placed at the points with position vectors $(2\mathbf{i} - \mathbf{j})$ m, $(2\mathbf{i} + 5\mathbf{j})$ m and $(4\mathbf{i} + 2\mathbf{j})$ m respectively. Find the position vector of the centre of mass of these particles.   (ULEAC)

11. A uniform ladder of length 4 m and mass 8 kg has its upper end resting against a smooth vertical wall and its lower end on rough horizontal ground. The ladder makes an angle of θ with the horizontal where $\tan\theta = 4$, and a man of mass 80 kg stands on the ladder at a distance of 3 m from the lower end. Given that the ladder is then in limiting equilibrium, find the coefficient of friction between the ladder and ground, giving your answer to two significant figures.   (WJEC)

12. A uniform beam AB of mass 20 kg and length 2 m is attached to a vertical wall by means of a smooth hinge at A. The beam is maintained in the horizontal position by means of a light inextensible string, one end of which is attached to the beam at B and the other end is attached to the wall at a point in the wall 2 m vertically above A.

(a) Calculate the tension in the string, in N to 3 significant figures.

A particle of mass M kg is now attached to the beam at B. Given that the string is about to break, and that the breaking tension of the string is 400 N,

(b) find the value of M to 3 significant figures.   (ULEAC)

13. The diagram shows a uniform rod AB, of length $2a$, in equilibrium in a vertical plane with the end A in contact with a vertical wall and the end B in contact with a horizontal floor. The normal reaction and frictional force at B are R and F respectively, acting in the directions shown. The corresponding forces at A, in the directions shown, are denoted by S and P, respectively. The rod is of weight W and is inclined at an angle θ to the horizontal.

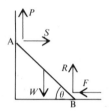

By resolving horizontally and taking moments about the centre of AB, or otherwise, express $R - P$ in terms of S and θ.

Also obtain an equation relating R, P and W and show that

$$R = \tfrac{1}{2}(W + 2S\tan\theta), \ P = \tfrac{1}{2}(W - 2S\tan\theta).$$

(a) Find, for the case when the wall is smooth and the coefficient of friction at B is 1, the value of $\tan\theta$ for limiting equilibrium.

(b) Show that, when the wall and floor are equally rough, the coefficient of friction being 1, limiting equilibrium is not possible for $\theta \neq 0$.   (AEB)

14. A uniform square lamina ABCD of side $5a$ is marked out into 25 equal small squares. The small square with one corner at C is removed. Find the distances from AB and AD of the centre of mass of the resulting lamina.   (WJEC)

15. A uniform lamina is in the shape of a rectangle ABCD where AD is of length 1 metre and AB is greater than AD. P is a point on AB such at AP = 1 m. The triangle PBC is removed and when suspended from P under gravity, the portion ADCP hangs with the side AD vertical. Show that PB is of length $\sqrt{3}\,m$ and find the distance of the centre of gravity of APCD from AP, giving your answer in its simplest surd form.   (WJEC)

16. A uniform rectangular plate OABC has mass $4m$, OA = BC = $2d$ and OC = AB = d. Particles of mass $2m$, m and $3m$ are attached at A, B, and C respectively on the plate. Find, in terms of d, the distance of the centre of mass of the loaded plate

(a) from OA   (b) from OC.

The corner O of the loaded plate, is freely hinged to a fixed point and the plate hangs at rest in equilibrium.

(c) Calculate, to the nearest degree, the angle between OC and the downward vertical.   (ULEAC)

The instruction for answering questions 17 to 20 is: if the following statement must *always* be true, write T, otherwise write F.

17. Three forces acting along the sides of a triangle are in equilibrium.

18. If a set of forces is in equilibrium the resultant moment about any two axes is zero.

19. The resultant moment of a set of forces is independent of the axis.

20. A uniform lamina in the form of a rectangle has one corner bent over as in the diagram. The centre of gravity of the resulting lamina lies on the diagonal AC.

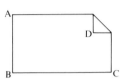

21. The diagram shows a uniform lamina in which

$PQ = PS = 2a$, $SR = a$, $S\hat{P}Q = R\hat{S}P = 90°$

The centre of mass of this lamina is G.

(a) Show that the distance of G from PS is $\frac{7}{9}a$.

(b) Find the distance of G from PQ.

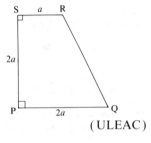

(ULEAC)

22.

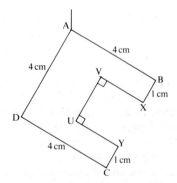

The diagram shows an ear-ring made from a uniform square lamina ABCD, which has each side of length 4 cm. Points X and Y are on the side BC and such that $BX = CY = 1$ cm. The square portion XYUV is removed and the resulting ear-ring is suspended from the corner A. The ear-ring hangs in equilibrium. The centre of mass of this ear-ring is G.

(a) State the distance, in cm, of G from AB.

(b) Find the distance, in cm, of G from AD.

(c) Find, to the nearest degree, the acute angle made by AD with the downward vertical. (ULEAC)

23.

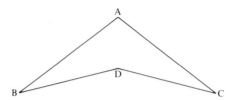

A uniform triangular lamina ABC has AB = AC = 5*a*, BC = 8*a* and D is the centre of mass of the lamina. The triangle BCD is removed from the lamina, leaving the plate ABDC shown in the diagram.

(a) Show that the distance of the centre of mass of the plate from A is $\frac{5a}{3}$.

The plate, which is of mass *M*, has a particle of mass *M* attached at B. The loaded plate is suspended from C and hangs in equilibrium.

(b) Prove that in this position CB makes an angle α with the vertical, where $\tan \alpha = \frac{1}{9}$.   (ULEAC)

24.

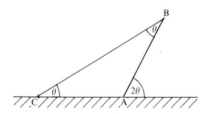

A uniform rod AB, of length 2*a* and weight *W*, is in equilibrium with the end A resting on a rough horizontal floor. The end B is connected to a point C on the floor by a light inextensible string and the plane ABC is vertical. The rod and the string make angles 2θ and θ, respectively, with the floor, as shown in the diagram. Show that the frictional force at A is

$$\tfrac{1}{2} W \cos 2\theta \cot \theta.$$

Given that the rod is on the point of slipping when θ = 30°, find the value of the coefficient of friction.   (JMB)

25. A uniform ladder of length 4*l* and mass *M* rests with one end A on rough horizontal ground and the other end B against a smooth vertical wall. The vertical plane containing AB is at right angles to the wall. The coefficient of friction between the ladder and the ground is $\frac{1}{5}$.

(a) A particle of mass 2*M* is attached to the ladder at C where AC = 3*l*. Given that equilibrium is limiting, show that the ladder is inclined at an angle α to the horizontal where $\tan \alpha = \frac{10}{3}$.

(b) The ladder is moved to a similar position where the wall is rough. The ladder again rests at an angle α to the horizontal and the coefficients of friction between the ladder and the wall and between the ladder and the ground are both μ ($\neq \frac{1}{5}$). The particle of mass 2*M* is moved to the top of the ladder. Given that equilibrium is limiting, show that $\mu^2 + 20\mu - 5 = 0$ (ULEAC)

26.

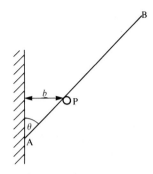

The diagram shows a uniform rod AB of length $2a$ and weight W, in equilibrium, resting on a rough peg P. The lower end A of the rod is in contact with a smooth vertical wall and the points A, P and B are in the same vertical plane perpendicular to the wall. The angle between AB and the wall is denoted by θ and b is the perpendicular distance of P from the wall. Find, in terms of W, a, b and θ, the reaction of the wall at A and the component normal to AB of the reaction of P. Deduce that equilibrium is only possible for $b \leqslant a \sin \theta$.
Show that the frictional force at P along AB, in the sense from A to B, is equal to

$$\frac{W(b - a \sin^3 \theta)}{b \cos \theta}.$$

Deduce that

$$a \sin^2 \theta (\sin \theta - \mu \cos \theta) \leqslant b \leqslant a \sin^2 \theta (\sin \theta + \mu \cos \theta),$$

where μ is the coefficient of friction.
Give a physical interpretation of the situation when b takes the lower value in the above inequality. (WJEC)

ANSWERS

CHAPTER 1

Exercise 1a – p. 5

1. (a) 6 m   (b) 2 m/s   (c) 1.5 m/s
    (d) 3 s   (e) 1.2 m/s

2.

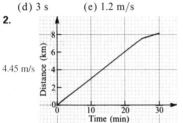

4.45 m/s

3. (a)

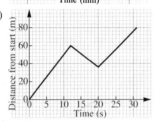

    (b) 5 m/s; 3 m/s; 4 m/s
    (c) (i) 4.2 m/s   (ii) 4.13 m/s (3 sf)

4. (a) 5.2 s
    (b) (i) 1.4 cm/s   (ii) 3.5 cm/s
    (c) (i) 1.3 cm/s   (ii) 2.6 cm/s

5.

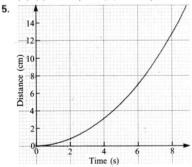

    (a) 0.8 m/s
    (b) (i) 1.8 m/s   (ii) 2.4 m/s
    (c) 5.5 s

6.

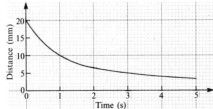

    (a) 1.6 mm/s   (b) 1.5 s
    (c) 3.3 mm/s

Exercise 1b – p. 12

1. (a) 8 mph/ minute   (b) 70 mph
    (c) 3 minutes
    (d) 4.3 mph/minute

2.

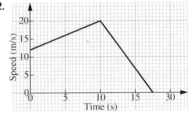

    (a) 0.8 m/s^2 (b) 2.5 m/s^2
    (c) 240 m

3. (a) (i) 20 s (ii) 15 s   (iii) 65 s
    (b) (i) 0.5 m/s^2 (ii) 0.67 m/s^2
    (c) 825 m

4. 15 s; 495 m

5. (a) (i) 17.6 m/s   (ii) 1.7 m/s
    (b) decelerating   (c) 4 m/s^2
    (d) 86 m

6.

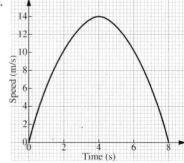

    (a) (i) ≈ 4 m/s^2 (ii) −3.5 m/s^2
    (b) 4 s
    (c) 75 m; an underestimate because the
    area of each trapezium is less than the
    area under the corresponding curve

7. (a) 26 km/min^2 (b) 23 km
    (c) 17 km

8. B

9. C

10. C

11. B

12. B

13. D

Exercise 1c – p. 18

1. (a),(d) and (f) are vectors
2. (a) (i) 2.3 m   (ii) 0   (iii) −1.2 m
   (b) (i) 3.4 m   (ii) 14 m
   (c) 0.08 m/s
3.

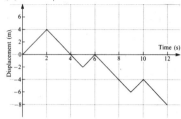

Exercise 1d – p. 20

1. (a) Correct
   (b) Incorrect; the direction is changing all
     the time
   (c) Incorrect; we do not know whether
     the speed is constant
2. (a) 1 m/s   (b) −2 m/s
   (c) 2 m/s   (d) −0.5 m/s
3. −4 m/s²
4. (a) −2 m/s²   (b) −3 m/s²
5. (a) 23 m/s   (b) −7 m/s
   (c) 5 m/s

Exercise 1e – p. 25

1. (a) (i)   15 m/s   (ii) −8 m/s   (iii) 0
   (b) (i) 0
     (ii) Ball hits wall and its direction is
      reversed
   (c) After 3 s
2.

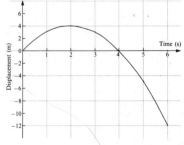

   (a) (i) 2 m/s   (ii) −2 m/s
     (iii) 0   (iv) −4 m/s
   (b) (i) 2 m/s   (ii) 2 m/s
     (iii) 2 m/s   (iv) 4 m/s
   (c) ≈ 4 m/s   (d) t = 2
3. (a) s   0   4   6   6   4   0   −6

   (b)

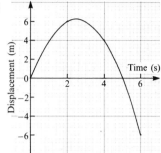

   (c) 2.5 s   (d) −5 m/s
   (e) (i) −1 m/s   (ii) 3 m/s
4. (a) 1 s, 1.75 s, 2.5 s, 3 s, 3.5 s, 3.8 s
   (b) (i) 1 m/s   (ii) 0.8 m/s   (iii) 0.7 m/s
   (c) (i) −1 m/s   (ii) 0   (iii) 0
   (d) −0.22 m/s
   (e) (i) −0.9 m/s   (ii) 0.5 m/s

Exercise 1f – p. 29

1. (a) 3 m/s²   (b) 1 m/s²
   (c) (i) 152 m   (ii) 400 m   (iii) 200 m;
     The car travels in the same direction
     all the time.
2.

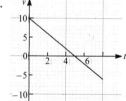

   360 m

3.

   (a) t = 5
   (b) 16 m
   (c) 34 m

4. (a)   (i) Accelerates at a reducing rate to
       zero acceleration
     (ii) Constant velocity
     (iii) Accelerates at varying rate
   (b) 185 m; less, because trapeziums have
     smaller area than that under the curve
5. (a) (i) after 8½ s   (ii) after 20 s
   (b) ≈ 1 m/s²; ≈ −0.5 m/s²
   (c) 6 s, 15 s after starting
   (d) The girl stops accelerating and begins
     to slow down.
6. C   **7. B**   **8. C**   **9. A**

CHAPTER 2

Exercise 2a – p. 38

1. 5
2. −1.2
3. 6
4. 60
5. 2
6. −2
7. 6
8. 1.5
9. −54
10. −8
11. No. He needs 85 m to come to rest and is travelling at 17 m/s at collision
12. 0.32 m/s²; 4.38 m/s
13. (a) 16 m   (b) 7 m
14. (a) $t = 2$ and 5   (b) 8.22 s
15. About 7 yards (0.0042 miles)
16. $-\frac{1}{30}$ m/s²
17. (a) 6.5 m/s   (b) −1.5 m/s²
    (c) 8.7 s
18. 48 m
19. 612 m
20. 12.4 s; 617 m
22. Yes, after 07.8 s

Exercise 2b – p. 44

1. (a) 123 m   (b) 11.0 m/s
2. (a) 30.6 m   (b) 5 s
3. (a) 4.9 m   (b) 19.6 m   (c) 24.5 m
4. (a) 11.0 m   (b) 14.7 m/s
5. (a) 3.19 s   (b) 31.3 m/s
6. (a) (i) 62.7 m/s   (ii) 62.5 m,/s
    (b) The value of g depends upon the distance from the centre of the earth and increases as the distance decreases. Because the earth is not a perfect sphere, this distance is smaller at the poles than at the equator.
7. 50.1 m
8. 28 m/s
9. (a) 1.12 s
    (b) 2.55 s; 18.0 m/s in each case
10. 18 m
11. (a) 11.5 m   (b) 2.63 s
12. 0.4 m
13. 440 m
14. 13 m/s; 2 s
15. 37.4 cubits/s

CHAPTER 3

Exercise 3a – p. 52

1. (a) $-\mathbf{a}+\mathbf{b}$ (b) $-\mathbf{b}+\mathbf{a}$
2. (a)   (b)

    (c)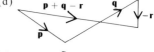

    (d)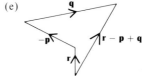

    (e)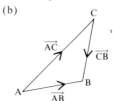

3. (a) $-\mathbf{p}+\mathbf{q}$ (b) $-\mathbf{q}+\mathbf{r}$
    (c) $-\mathbf{p}+\mathbf{r}$
4. (a) $-\mathbf{a}+\mathbf{b}$ (b) $\frac{1}{2}(\mathbf{b}-\mathbf{a})$
    (c) $\frac{1}{2}(\mathbf{a}-\mathbf{b})$
5. (a) $\mathbf{a}+\mathbf{b}$ (b) $\mathbf{b}+\mathbf{c}$
    (c) $-(\mathbf{b}+\mathbf{c})$ (d) $-(\mathbf{a}+\mathbf{b}+\mathbf{c})$
6. (a)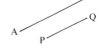

    (b)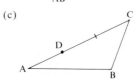

    (c)
    A ———————— B
    D
    C

8. (a) $\mathbf{b}$ (b) $-\mathbf{b}$ (c) $2(\mathbf{a}-\mathbf{b})$
10. (a) (i) $\overrightarrow{AD}$ (ii) $\overrightarrow{AC}$ (iii) $\overrightarrow{EB}$
    (b) $\overrightarrow{AE}+\overrightarrow{ED}$, $\overrightarrow{AB}+\overrightarrow{BC}+\overrightarrow{CD}$,
    $\overrightarrow{AB}+\overrightarrow{BD}$, etc
11. 2

12. (a) $\overrightarrow{BE} = \mathbf{b} - 2\mathbf{a}$, $\overrightarrow{CF} = \mathbf{a} - 2\mathbf{b}$
(b) $h = \frac{1}{3} = k$
(c) $2:1$

Exercise 3b – p. 58
1. 13 km, 157°
2. 13 km, 157°
3. 25 m/s, 016°
4. 20 N, 217°
5. 20.3 m, 146°
6. 7.81 m/s, 065°
7. 96.4 N, 39° to 40 N force
8. 5.2 m/s, 30° to 6 m/s
9. 14.9 m, 314°
10. 10.1 N, 54° anticlockwise from 10 N
11. 4.95 km, 046°
12. 444 km/h, 005°
13. 1100 m, 103°; 283°
14. Bearing 233°; 5 km/h
15.

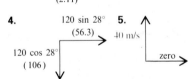

2.65 km/h

16. 166 km/h, 102°

Exercise 3c – p. 62
1. (a) 6 sin 25°   (b) 10 sin 20°
(c) 52 cos 20°   (d) 20 cos 60°
(e) 2 cos 50°   (f) 8 sin 40°

2.

5 sin 65°
(4.53)
5 cos 65°
(2.11)

3.

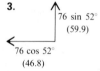

76 sin 52°
(59.9)
76 cos 52°
(46.8)

4.

120 sin 28°
(56.3)
120 cos 28°
(106)

5.

40 m/s
zero

6.

zero
10 m/s

7. 30 sin 70°
(28.2)
30 cos 70°
(10.3)

8.
20 sin 30°
(10)
20 cos 30°
(17.3)

9.
5 cos 60°
(2.5)
5 sin 60°
(4.33)

10.

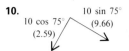

10 cos 75°
(2.59)
10 sin 75°
(9.66)

11.

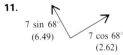

7 sin 68°
(6.49)
7 cos 68°
(2.62)

12.

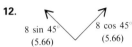

8 sin 45°
(5.66)
8 cos 45°
(5.66)

13.

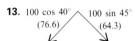

100 cos 40°
(76.6)
100 sin 45°
(64.3)

14.

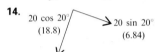

20 cos 20°
(18.8)
20 sin 20°
(6.84)

15.

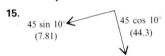

45 sin 10°
(7.81)
45 cos 10°
(44.3)

16.

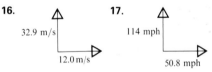

32.9 m/s
12.0 m/s

17.
114 mph
50.8 mph

17. 50.8 mph parallel
114 mph perpendicular
18. 68.8 m/s parallel
162 m/s perpendicular
19. 80.6 N parallel
59.1 N perpendicular

Exercise 3d – p. 67
1. (a) $4\mathbf{i} + 3\mathbf{j}$ (b) $4\mathbf{j}$
(c) $-2\mathbf{i} + 4\mathbf{j}$ (d) $-5\mathbf{i}$
(e) $-4\mathbf{j}$ (f) $-2\mathbf{i} - 3\mathbf{j}$
2. (a) 5   (b) 4   (c) $\sqrt{20}$
(d) 5   (e) 4   (f) $\sqrt{13}$
3. (a) $12.1\mathbf{i} + 7\mathbf{j}$ (b) $-20\mathbf{j}$
(c) $-53\mathbf{i} - 53\mathbf{j}$ (d) $240\mathbf{i} + 320\mathbf{j}$
4. (a) $9\mathbf{i} - 4\mathbf{j}$

(b) $3\mathbf{i} + 6\mathbf{j}$

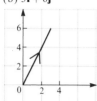

5. (a) $\mathbf{i} - \mathbf{j}$ (b) $\mathbf{i} + \mathbf{j}$

6. (a) $-\mathbf{i} - 2\mathbf{j}$ (b) $-3\mathbf{i}$

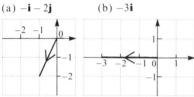

7. (a) 5   (b) 13   (c) $\sqrt{2}$
8. $24\mathbf{i} - 18\mathbf{j}$
9. (a) $-11\mathbf{i} + 4\mathbf{j}$ (b) $\sqrt{137}$
10. $8\mathbf{i} - 6\mathbf{j}$ or $-8\mathbf{i} + 6\mathbf{j}$
11. (a) $-35\mathbf{i}$ (b) $8\mathbf{i} - 4\mathbf{j}$

CHAPTER 4
Exercise 4a – p. 75

1.

2.
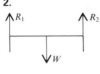

3.

4.

5.

6.

7.
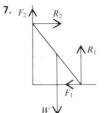

8. (a) (b) (c)

9. (a) (b)   **10.**
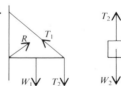

11.
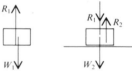

12.
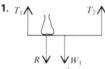

13.
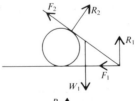

14.
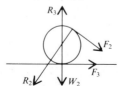

15. (a) 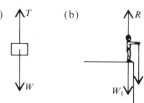 (b)

10. (a)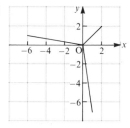

Exercise 4b – p. 82

1. 717 N, 33°
2. 1.43 N, 170°
3. 22.9 N, 234°
4. 12.1 N, 38°
5. 8.28 N, 51°
6. 17.2 N, 63°
7. (a)

(b) $-3\mathbf{i} - 4\mathbf{j}$
(c) 5 N, 127°

11. 8.22 N, 49°
12. 261 N, bearing 305°
13. 292 N, 6°
14. $9\mathbf{i} + 5\mathbf{j}$
15. $(3 - 2\sqrt{2})\mathbf{i} + (5 - 2\sqrt{2})\mathbf{j}$; 2.18 m/s
16. 7.08 km/h, 133°
17. (a)

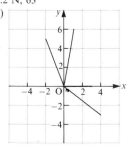

(b) $6\mathbf{i} + 8\mathbf{j}$ (c) 10 N, 53°

8. (a)

(b) $0.364\mathbf{i} + 7.71\mathbf{j}$ (c) 003°
(d) 7.71 m/s
18. 1.01 N, 83°

CONSOLIDATION A

Miscellaneous Exercise A – p. 87

1. B
2. C
3. A
4. $3\mathbf{i} + 7\mathbf{j}$
5. B
6. $-2\mathbf{i} + 3\mathbf{j}$; 124°
7. $\frac{2}{9}\mathbf{a}$
8. (a) $9\mathbf{i} + 17\mathbf{j}$ (b) 19.2 N
   (c) 0.468 $(9/\sqrt{370})$

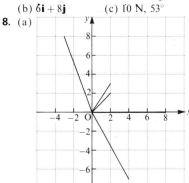

(b) $12\mathbf{i} + 5\mathbf{j}$ (c) 13 m/s, 23°

9. (a)

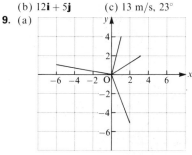

9. The midpoint of XY
10.

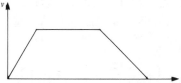

(a) 40 km/h
(b) 12 minutes   (c) 16 km

11. $s = 19.6(t - 2) - 4.9(t - 2)^2$,
   $s = 19.6t - 4.9t^2$; 2, 9.8 m/s, 9.8 m/s

(b) $2\mathbf{j}$ (c) 2 m, 90°

12. OA : 0.25 m/s², AB : 0, BC : -0.5 m/s²;
8000 m

13. (a) 1.875 m/s²   (b) $\frac{1}{2}$

14.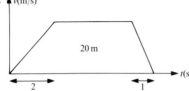

15. (a) 50 s   (b) 24.2 m/s

16. (a)

(b) 72 km/h   (c) 2160 km/h²

17. (a) 5.55 km   (b) 2.12 minutes

18. $(k-4)\mathbf{i}+4\mathbf{j}$; $\sqrt{(k^2-8k+32)}$; 1, 7; $\frac{4}{3}$

19. $4v$, v, $v^2-60v+800=0$, $v=20$
(for $v=40$, it takes 160 s to accelerate,
which is greater than the total time, so
$v \neq 40$)

20. $\overrightarrow{OC}=2\mathbf{i}+14\mathbf{j}$, $\overrightarrow{OD}=6\mathbf{i}+12\mathbf{j}$

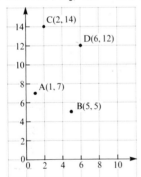

21. F   **22.** F   **23.** F   **24.** F

25. (b) $\overrightarrow{PN}=\mathbf{q}-2\mathbf{p}$, $\overrightarrow{QM}=\mathbf{p}-2\mathbf{q}$
(d) $2\mu\mathbf{q}+(1-\mu)\mathbf{p}$

26.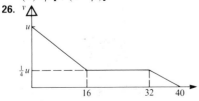

(a) $\frac{3}{64}u$ m/s²   (b) $\frac{1}{32}u$ m/s²
(c) 3 m/s

27.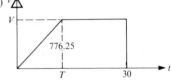

(a) 900 m   (b) 1

28. (a)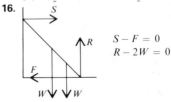

(b) $V=2.3T$

29. 2

EXERCISE 5

Exercise 5a – p. 94

1. $P=10$

2. $P=20$, $Q=90$

3. $P=12$, $Q=4$

4. $P=36$, $Q=28$

5. $P=18$

6. $P=12$, $Q=26$

7. Yes, horizontal

8. Yes, vertical

9. No

10. Yes, neither vertical nor horizontal

11. No

12. Yes, neither horizontal nor vertical

13. (a)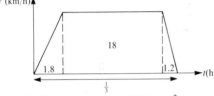

(b) (i) $T>F$ (ii) $T=R$

14. Yes $\parallel$ to $\mathbf{j}$

15. (a) $R\cos 30° + N - W$
(b) $R\sin 30° - F$
(c) $R\frac{\sqrt{3}}{2}+N-W=0$, $\frac{1}{2}R-F=0$

16.

$S-F=0$
$R-2W=0$

17.

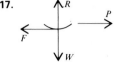

(a) $P = F$ (b) $R = W$

4.

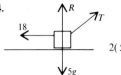

$2(5g + 18) = 134$ N

5. 41 m

Exercise 5b – p. 99

1. 1.5 m/s^2
2. 28 N
3. 40 kg
4. $\dfrac{5\mathbf{i} - 12\mathbf{j}}{26}$, $\frac{1}{2}$ m/s^2
5. $5(7\mathbf{i} + 2\mathbf{j}) = 35\mathbf{i} + 10\mathbf{j}$ (in newtons)
6. (a) 38 N
  (b) $P = 10$ N, $Q = 8$ N
  (c) $P = 34$ N, $Q = 30$ N
7. (a) 6 m/s$^2 \rightarrow$ (b) 16 m/s$^2 \rightarrow$
  (c) $\frac{20}{3}$ m/s$^2 \rightarrow$
8. (a) $m = 4$, $P = 23$ N
  (b) $m = 5$, $P = 30$
  (c) $m = 10$, $P = 40$
9. $P = 8$, $Q = 31$
10. 8 N
11. $156\frac{1}{4}$ m
12. 30 N
13. (a) $-0.4\mathbf{i} + \mathbf{j}$ (b) $-0.8\mathbf{i} + 2\mathbf{j}$
14. 16 kg
15. 100 m
16. 14 m/s

Exercise 5c – p. 102

1. (a) 49 N   (b) 15 kg   (c) 0.59 N
2. (a) 6 N   (b) 122.5 kg   (c) 0.072 N
3. 31.2 N
4. (a) 28.8 N   (b) 101 N   (c) 58.8 N
5. (a) 5.47 kg   (b) 87.5 kg   (c) 7.14 kg
  (d) 7.14 kg
6. (a) (i) 18 000 N   (ii) 6790 N
  (b) 750 kg
7. 10 920 N
8. (a) 59.2 N   (b) 19.2 N

Exercise 5d – p. 106

1. (a) $(g + 1)$ N   (b) $(g + 5)$ N
2. (a) 583 kg   (b) 5070 N
3. (a)   (b)   (c)

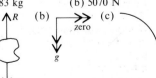

Exercise 5e – p. 109

1.

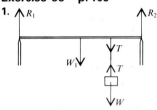

2.

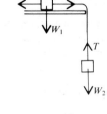

3.

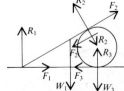

4.

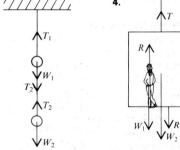

5.

6.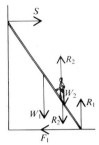

Exercise 5f – p. 116

1. (a) $a = g/2$, $T = 3g$
   (b) $a = g/5$, $T = 80g/7$
   (c) $a = \left(\dfrac{M-m}{M+m}\right)g$, $T = \dfrac{2mg}{M+m}$
2. (a) $a = g/3$, $T = 20g/3$
   (b) $a = g/5$, $T = 48g/5$
   (c) $a = g/3$, $T = 4g/3$
3. (a) $(\sqrt{3}-1)g/3$ (b) $2(\sqrt{3}+2)g/3$
4. (a) $g/3$ (b) $25g/6$
5. (a) $3g/8$ (b) $15g/8$
6. $a = 2$ m/s², $T = 600$ N
7. (a) (i) $35(2g+1)$ N; a particle or a
               small block   (ii) $435(2g+1)$ N
   (b) (i) $35(2g-1)$ N   (ii) $435(2g-1)$ N
8. $g(\sqrt{2}+\sqrt{6})$ N, bisecting angle between faces
9. g metres; g m/s
10. (a) $g/6$
    (b) $T_1 = 10mg/3$, $T_2 = 7mg/3$
11. (a) $1885g/3$, $260g/3$
    (b) $580g$, $80g$ (c) $522g$, $72g$

Exercise 5g – p. 119

1. (a) $g/9$ (b) $\sqrt{2g}/3$
   (c) $\sqrt{2}/3\sqrt{g}$
2. (a) $g/3$ (b) $g/18$
3. (a) $\sqrt{g/2}$ (b) g N
4. (a) g (b) 0
5. (a) 3.3 m/s²   (b) 2.8 m/s
6. (a) freely under gravity
   (b) 1.25 m/s
7. (a) 1480 N   (b) 3330 N
8. (a) 3.08 m/s²   (b) -0.25 m/s²
   (c) 80 s

CHAPTER 6

Exercise 6a – p. 125

1. (a) $P = 30\sqrt{3}$, $Q = 30$
   (b) $P = 12$, $Q = 12\sqrt{3}$
   (c) $P = 100 \sin 20° = 34.2$,
       $Q = 100 \cos 20° = 94.0$
2. (a) $P = 24$, $Q = 22\sqrt{3}$
   (b) $P = 24$, $Q = 15\sqrt{3}$
   (c) $P = 8$, $\theta = 60°$
3. (a) $P = 13\sqrt{3}$, $Q = 0$
   (b) $\theta = 60°$, $P = 4\sqrt{3}$
   (c) $P = 10$, $Q = 8$
4. $5g$ N
5. $90°$; $T_1 = 12$ N, $T_2 = 16$ N
6. $w/\sqrt{3}$ $(w\sqrt{3}/3)$
7. $w/2$
8. $w/2$
9. (a) $30°$ (b) $18°$
10. (anticlockwise from P) $P\mathbf{i}$, $Q\mathbf{j}$,
    $\dfrac{5\sqrt{3}}{2}\mathbf{i} - \dfrac{5}{2}\mathbf{j}$, $\dfrac{-3\sqrt{3}}{2}\mathbf{i} - \dfrac{3}{2}\mathbf{j}$, $2\sqrt{3}\mathbf{i} - 6\mathbf{j}$;
    $P = -3\sqrt{3}$, $Q = 10$
11. (a) $q = -16$, $p = 0$
    (b) $p = 4$, $q = -7$
12. (a) $a = 7$, $b = -11$ (b) $-11\mathbf{i} + 2\mathbf{j}$

Exercise 6b – p. 132

1. $\sqrt{3}/3$ (0.577)
2. $\frac{1}{3}$
3. $6(2+\sqrt{3})$ N $(22.4$ N$)$
4. $\dfrac{1}{3-\sqrt{3}}$ (0.789)
5. $12(1+\sqrt{3}/5)$ N (16.2)
6. 0.309
7. $\frac{1}{2}$
8. (a) 17.4 N   (b) 9.18 N
9. (b) $11w/23$
   (c) $\left(\dfrac{2+5\sqrt{3}}{5-2\sqrt{3}}\right)w$ $(\approx 6.94w)$
10. (a) 10 N, no
    (b) 40 N, no (just on the point of slipping)
    (c) 40 N, yes
11. $3g/10$; $7Mg/10$
12. $g\sqrt{2}/15$; $2g\sqrt{2}/3$

CHAPTER 7

Exercise 7a – p. 139

1. (anticlockwise from 20 N)
   $20\sqrt{3}$ J, 0, -24 J, 0
2. (anticlockwise from 8 N)
   -24 J, 0, 48 J
3. (anticlockwise from 7 N)
   28 J, 0, -8 J, -16 J
4. 63 J
5. 6100 J
6. 3500 J

7. 24 kJ

8. 110 kJ

9. (a) 180 J   (b) 290 J

10. 2800 kJ

11. (a) 1100 J   (b) 8300 J

12. −4400 J

13. 1400 J

14. 25 N

15. 100 J

16. $\frac{1}{4}$

17. 11 kJ; 14 kJ

18. (a) 270 N   (b) 170 N
(c) 250 J   (d) 930 J

19. 7500 J   (a) 500 N↓
(b) 1300 N↓   (c) 12 kJ

20. (a) $\frac{1}{4}g$ N   (b) $\frac{1}{2}g$ N   (c) $\frac{1}{2}g$ N
(d) $\frac{1}{2}g$ J   (e) $\frac{1}{2}g$ J

Exercise 7b – p. 146

1. 270 W

2. 2 kW

3. 18 W

4. 4.8 W

5. 8.6 W

6. (a) 24 kW   (b) 1800 N
(c) 20 m/s

7. 1000 kW

8. 1400 N

9. 360 W; 4.6 m/s

10. (a) 17 m/s
(b) 40 m/s; e.g. resistance unlikely to be constant in different conditions

11. (a) 31 m/s   (b) 23 m/s
(c) 26 m/s

12. (a) 29 m/s   (b) 40%

13. (a) 96 kW   (b) 6 kN
(c) 16 m/s

14. (a) 500/16   (b) 33 m/s   (c) 2°

Exercise 7c – p. 150

1. (a) 38 kW   (b) 0.61 m/s²

2. 0.65 m/s²

3. 2.2 m/s²

4. 0.13 m/s²

5. (a) 30 kW   (b) 1.2 m/s²

6. (a) 19.2   (b) 0.48 m/s²

7. (a) 14 m/s   (b) 0.55 m/s²

8. (a) 310 N   (b) 0.5 m/s²

9. (a) $R = 1000$, $H = 21$
(b) 0.25 m/s²

10. 1.8 m/s²

CHAPTER 8

Exercise 8a – p. 156

1. (a) 431   (b) 0.816

2. (a) 324 J   (b) 60 000 J
(c) 2000 J   (d) 65 J

3. 14 J, 4.24 m/s, 22 kg

4. (a) (i) 0   (ii) 4200 J
(b) 4800 J   (c) 1900 J   (d) 1400 J

5. (a) 144 J   (b) 598 J

6. (a) 79 kJ   (b) 25 m/s (90 km/h)

7. (a) 10 m/s   (b) 25 m

8. (a) 5300 J   (b) 3200 J

9. (a) 0.48 m³   (b) 480 kg
(c) 38.9 kJ

Exercise 8b – p. 160

1. 22.9 N

2. 576 J

3. 1.64 N; 6.41 m/s

4. 10.6 J; 6.22 N

5. 12.4 m/s

6. (a) 6.20 m/s   (b) 4.77 m/s

7. 21.6 kN

8. (a) 137 J   (b) 148 J

9. (a) 16.1 m   (b) 2.99 m/s

10. 26.1 m/s

11. 3.4 N

12. (a) 0.5 m³   (b) 4900 N
(c) 64.2 kJ

13. (a) 2   (b) 10.2 kW

14. 3.76 m/s

Exercise 8c – p. 165

1. 1.84 m

2. 9.86 m/s

3. (a) 4.27 m   (b) 6.05 m/s
(c) 9.14 m/s

4. 1.63 m

5. 4.85 m/s

6. 3.93 m/s

7. 1.89 m

8. 3.13 m/s

9. 0.3 m

10. 5.6 m/s

11. $\sqrt{gl}$

12. $\sqrt{2gl/3}$

CONSOLIDATION B

Miscellaneous Exercise B – p. 169

1. $P \cos \theta = 4$, $P \sin \theta = \sqrt{3}$,
$\tan \theta = \frac{1}{4}\sqrt{3}$, $P = \sqrt{19}$

2. (a) 57.3 N   (b) 53.9 N

3. 256 s, 7440 m

4. $\dfrac{u^2}{g}, \dfrac{2u^2}{3g}$

5. (a) 15 m/s   (b) 112 N

6. (a) 4.4 N in directioon of Ox
     (i) 4.4 N in direction of xO
     (ii) 44 m/s^2
     (b) 22.8 N at $-21.6°$ to PA

7. (a) 2.7 N   (b) 10.6 N

8. 1.4 m/s^2, 0.672 N

9. $\sqrt{2gl/3}$

10. (a) 1.96 m/s^2 (b) 0.784 N
     (c) If the string breaks while B is on the
     table, B moves on at constant speed
     to the edge; A moves down with
     acceleration g.

11. (b) 0.632 s, 3.79 m/s   (c) 3.12 m

12. (a) 0.61 m/s^2 (b) 27.57 N

13. $a = \tfrac{1}{3}g$, $T = \tfrac{4}{3}mg$

14. C   **15.** A   **16.** B   **17.** C

18. (a) 20 m/s   (b) 11.44 m/s

19. 700 N, 38.2°; 1400 J; 140 kg

20. 9 770 000 J

21. (a) 11.2 m   (b) 9.5 J

22. (a) 11 000 J   (b) 122 N
     (c) 32.1 m

23. 20 N

24. T   **25.** F   **26.** T   **27.** T
28. T   **29.** T   **30.** F   **31.** T

32. 0.164 m/s^2

33. 350 kW   (a) 400 s   (b) 8 km

34. (b) 17 000 J

35. (b) 18 m/s

36. (a) 0.25 m/s^2
     (b) 22.5 m/s; 12.5 s; 187.5 m

37. 4000 kg, 0.25

38. (b) 20 s   (c) 884 W   (d) 0.5 m/s^2

39. (a) 2000 J; 2000 J;
     $2\sqrt{10}$ m/s (6.32 m/s)
     (b) 40.1 m/s; 64 360 J

CHAPTER 9

Exercise 9a – p. 184

1. (a) 21 m/s ↗   (b) 23 m/s ↘

2. (a) 12 m   (b) 15 m
     (c) 7 m

3. (a) 25 m/s; (24, 13)
     (b) 24 m/s; (48, 16)

4. 1 s; 3.7 m

5. (a) $20\sqrt{2}$ m/s (28 m/s, 2 sf)
     (b) $56\sqrt{2}$ m (79 m, 2 sf)

6. $20\mathbf{i} + 10\mathbf{j}$

7. $x = 4.33$, $y = 1.25$; 4.5 m

8. $9.6\mathbf{i} - 11.5\mathbf{j}$

9. (a) $\mathbf{v} = \mathbf{i} + (2 - 10t)\mathbf{j}$;
     $\mathbf{r} = t\mathbf{i} + (2t - 5t^2)\mathbf{j}$
     (b) $\mathbf{v} = \mathbf{i} - 13\mathbf{j}$; $\mathbf{r} = 1.5\mathbf{i} - 8.3\mathbf{j}$

10. (a) $\mathbf{v} = 5\mathbf{i} + (3 - gt)\mathbf{j}$;
     $\mathbf{r} = 5t\mathbf{i} + (3t - \tfrac{1}{2}gt^2)\mathbf{j}$
     (b) $\mathbf{v} = 4\mathbf{i} - (1 + gt)\mathbf{j}$;
     $\mathbf{r} = 4t\mathbf{i} - (t + \tfrac{1}{2}gt^2)\mathbf{j}$
     (c) $\mathbf{v} = 10\mathbf{i} + (10\sqrt{3} + gt)\mathbf{j}$;
     $\mathbf{r} = 10t\mathbf{i} + (10\sqrt{3}t - \tfrac{1}{2}gt^2)\mathbf{j}$
     (d) $\mathbf{v} = 10\mathbf{i} + (20 - gt)\mathbf{j}$;
     $\mathbf{r} = 10t\mathbf{i} + (20t - \tfrac{1}{2}gt^2)\mathbf{j}$

11. (a) $t = 0.8$, $t = 4$ (b) 32 m

12. (a) 15 m/s, 9.5° below →
     (b) 10 m/s, 14° below →

13. $20/g$ s; $200/g$ m

14. $\tfrac{8}{3}\mathbf{i} + \tfrac{49}{6}\mathbf{j}$

15. (a) 2.6
     (b) No, the ball lands 24 m from the
     wall

16. $a = 40$, $b = 110$

17. 2.6 s

18. 39 m

19. Yes by almost 1 m

20. (a) 35 m   (b) 215 m

21. 63°

22. 54 m

23. (a) $u\mathbf{i} + u\mathbf{j}$
     (b) (i) $u\mathbf{i} + (u - 20t)\mathbf{j}$
     (ii) $u\mathbf{i} + (u - 20)\mathbf{j}$
     (c) $2u\mathbf{i} + (2u - 20)\mathbf{j}$

24. (a) 3.5 s   (b) 1.5 s

25. (a) $V\cos\theta$ (b) $V\sin\theta + gt$
     (c) $Vt\cos\theta$ (d) $Vt\sin\theta + \tfrac{1}{2}gt^2$

26. (a) 35 m/s   (b) 53 m

27. Between 4.7 m and 15 m

Exercise 9b – p. 196

1. 7.2 m; 50 m

2. 22° or 68°

3. 30 m/s; 40 m

4. 26 m/s; 26 m

5. 90 m

6. 27°; 130 m

7. (a) 48 m   (b) 52 m

8. $y = 0.839x - 0.003\,41x^2$; 15 m

9. 12°, 5.9 m

10. (a) $y = \tfrac{4}{3}x - \tfrac{5}{9}x^2$
     (b) $y = -x - \tfrac{1}{90}x^2$

11. 1.9 s; 25 m

12. 31°

13. (a) $h\sqrt{3}$ (b) $2h$ (c) $h\sqrt{3}$

14. $15°$

15. $45°$ to $53°$

16. (a) $\frac{11}{8}h$ (b) $\frac{3}{2}h$ (c) $\frac{3}{8}h$
   (d) $-2h$

17. $74°$ (not $18°$ as this is not a "skyer")

18. $45°$; 160 m/s

19. $y = 1.4x - 0.2x^2$

20. No; hits net 0.14 m below cord

21. First cleared by 3.2 m; second one is just hit

22. (a) 3 s   (b) 16 m   (c) $42°$

Exercise 9c – p. 202

1. $36\mathbf{i} + 8\mathbf{j}$

2. 2.9 s

3. (a) $37°$ (b) 100 m

4. (b) $V^2\sqrt{3}/g$

5. (b) 1 s after Q is projected
   (c) P at $14°$ below $\rightarrow$; Q at $19°$ above $\rightarrow$

6. (a) 15 m   (b) $\sqrt{3}$ s (1.7 s)

7. (a) 45 m   (b) 3 s

8. 50 m

CHAPTER 10

Exercise 10a – p. 210

1. 85 m s^{-1}

2. 50 m s^{-2}

3. 26 m s^{-2}

4. $82\frac{2}{3}$ m

5. 17 m s^{-1}; 12 m

6. $t = 2$ and 4

7. $t = \frac{2}{3}$ and $\frac{1}{3}$

8. $v = t^2$; $r = \frac{1}{3}t^3$

9. $29\frac{1}{3}$ m s^{-1}; $47\frac{1}{3}$ m

10. (a) $t = 3.1$, $t = -0.43$
   (b) 21 m, -0.90 m

11. $v = 3(t + \frac{1}{3})^2 + \frac{35}{3}$ which is always positive

12. (a) $31\frac{1}{2}$ m s^{-1} (b) 27 m

13. (a) $\frac{1}{2}(3t^2 + 1)$ (b) 10 m s^{-1}

14. (a) 2.53 m s^{-1} (b) 2.5 m s^{-1}

15. (a) $10\frac{2}{3}$ m s^{-1} (b) $t = 2$
   (c) $13\frac{1}{3}$ m

16. $s = \dfrac{32t^2}{15(t+1)}$

17. $v = u + at$, $s = ut + \frac{1}{2}at^2$

18. (a) 8 m s^{-1} (b) $5\frac{1}{3}$ m
   (c) $2\sqrt{\frac{7}{3}}$ (3.06)   (d) 9.85 m

19. (a) $2\left(1 + \dfrac{1}{t^3}\right)$ (b) 6.048 N

20. $\frac{1}{2}t^3 + \frac{5}{2}t^2 + 6t - 26$

21. $66\frac{2}{3}$ m

Exercise 10b – p. 216

1. (a) (i) $\mathbf{v} = 6t^2\mathbf{i} + 6t\mathbf{j}$, $\mathbf{a} = 12t\mathbf{i} + 6\mathbf{j}$
    (ii) When $t = 2$, $\mathbf{v} = 24\mathbf{i} + 12\mathbf{j}$,
                         $\mathbf{a} = 24\mathbf{i} + 6\mathbf{j}$
       When $t = 3$, $\mathbf{v} = 54\mathbf{i} + 18\mathbf{j}$,
                           $\mathbf{a} = 36\mathbf{i} + 6\mathbf{j}$

   (b) (i) $\mathbf{v} = (2t + 1)\mathbf{i} - 2t\mathbf{j}$, $\mathbf{a} = 2\mathbf{i} - 2\mathbf{j}$
    (ii) When $t = 1$, $\mathbf{v} = 3\mathbf{i} - 2\mathbf{j}$,
                         $\mathbf{a} = 2\mathbf{i} - 2\mathbf{j}$
       When $t = 4$, $\mathbf{v} = 9\mathbf{i} - 8\mathbf{j}$,
                         $\mathbf{a} = 2\mathbf{i} - 2\mathbf{j}$

   (c) (i) $\mathbf{v} = 3\mathbf{i} + 12(t^2 + 1)\mathbf{j}$, $\mathbf{a} = 24t\mathbf{j}$
    (ii) When $t = 1$, $\mathbf{v} = 3\mathbf{i} + 24\mathbf{j}$,
                         $\mathbf{a} = 24\mathbf{j}$
       When $t = 3$, $\mathbf{v} = 3\mathbf{i} + 120\mathbf{j}$,
                         $\mathbf{a} = 72\mathbf{j}$

   (d) (i) $\mathbf{v} = -\dfrac{2}{t^2}\mathbf{i} - \dfrac{12}{t^3}\mathbf{j}$, $\mathbf{a} = \dfrac{4}{t^3}\mathbf{i} + \dfrac{36}{t^4}\mathbf{j}$
    (ii) When $t = 1$, $\mathbf{v} = -2\mathbf{i} - 12\mathbf{j}$,
                        $\mathbf{a} = 4\mathbf{i} + 36\mathbf{j}$
       When $t = 2$, $\mathbf{v} = -\frac{1}{2}\mathbf{i} - \frac{3}{2}\mathbf{j}$,
                        $\mathbf{a} = \frac{1}{2}\mathbf{i} + \frac{9}{4}\mathbf{j}$

   (e) (i) $\mathbf{v} = \left(1 + \dfrac{1}{t^2}\right)\mathbf{i} + \left(2t + \dfrac{2}{t^3}\right)\mathbf{j}$,
        $\mathbf{a} = -\dfrac{2}{t^3}\mathbf{i} + \left(2 - \dfrac{6}{t^4}\right)\mathbf{j}$
    (ii) When $t = 1$, $\mathbf{v} = 2\mathbf{i} + 4\mathbf{j}$,
                        $\mathbf{a} = -2\mathbf{i} - 4\mathbf{j}$
       When $t = 3$, $\mathbf{v} = \frac{10}{9}\mathbf{i} + \frac{164}{27}\mathbf{j}$,
                        $\mathbf{a} = -\frac{2}{27}\mathbf{i} + \frac{52}{27}\mathbf{j}$

2. (a) $4y = 3x^2$ (b) $(y + 4)^2 = 16x$
   (c) $xy = 6$

3. (a) $2 - t$; 2, 1, 0
   (b) $\dfrac{3}{4 - 10t}$; $\frac{3}{4}$, -3, $-\frac{3}{4}$
   (c) $\frac{3}{2}t$; $\frac{1}{2}$, 1, $\frac{3}{2}$

4. $\mathbf{v} = t\mathbf{i} - 2t\mathbf{j}$; $\mathbf{r} = (\frac{1}{2}t^2 + 3)\mathbf{i} + (1 - t^2)\mathbf{j}$

5. (a) $-\mathbf{j}$ (b) $18\mathbf{i} + \mathbf{j}$ (c) $8\mathbf{i}$

6. $\mathbf{r} = 16\mathbf{i} + 8\mathbf{j}$; $8\sqrt{5}$ m s^{-1}

7. $\mathbf{v} = t(3\mathbf{i} - 2\mathbf{j})$; $3\sqrt{13}$ m s^{-1}

8. $2\sqrt{10}$ m s^{-2}

9. $\mathbf{v} = 9\mathbf{i} + 14\mathbf{j}$; $\mathbf{r} = 9\mathbf{i} + \frac{57}{2}\mathbf{j}$

10. (a) (i) $\mathbf{i}$ (ii) $2(\mathbf{i} - t\mathbf{j})$
   (b) The direction of motion always has a horizontal component so the direction can never be vertical.
   (c) 2 m s^{-2} ($-2\mathbf{j}$)

11. (a) (i) $\mathbf{v} = -gt\mathbf{j} + V\cos\alpha\,\mathbf{i} + V\sin\alpha\,\mathbf{j}$
   (ii) $\mathbf{r} = -\frac{1}{2}gt^2\mathbf{j} + Vt\cos\alpha\,\mathbf{i}$
      $+ Vt\sin\alpha\,\mathbf{j}$
   (b) $y = x\tan\alpha - \dfrac{gx^2}{2V^2\cos^2\alpha}$

12. (a) (i) $6\mathbf{i}$ (ii) $4\mathbf{j}$
   (iii) $\mathbf{v} = 6\mathbf{i} + 4t\mathbf{j}$, $\mathbf{r} = 6t\mathbf{i} + 2t^2\mathbf{j}$
   (b) $18y = x^2$

13. (a) $8\mathbf{i} + 4\mathbf{j}$; $4\mathbf{i} + 2\mathbf{j}$
   (b) $\mathbf{r} = 2t^2\mathbf{i} + (t^3 + 3)\mathbf{j}$

14. (a) $\mathbf{F}_1 = 8\mathbf{i} + 6\mathbf{j}$, $\mathbf{F}_2 = 6\mathbf{i} + 3\mathbf{j}$,
      $\mathbf{F}_1 + \mathbf{F}_2 = 14\mathbf{i} + 9\mathbf{j}$
   (b) $\mathbf{a} = 7\mathbf{i} + \frac{9}{2}\mathbf{j}$, $\mathbf{v} = 7t\mathbf{i} + \frac{9}{2}t\mathbf{j}$
   (c) $14\mathbf{i} + 9\mathbf{j}$

15. (a) $2\mathbf{i} + \frac{3}{2}\mathbf{j}$ (b) 6.25 J

16. (a) $\mathbf{v} = 6\mathbf{i} - 8t\mathbf{j}$, $\mathbf{a} = -8\mathbf{j}$
   (b) $\mathbf{F} = -32\mathbf{j}$ (c) 2120 J
   (d) 2048 J

17. (a) $2\mathbf{i} - 3\mathbf{j}$ (b) $t(2\mathbf{i} - 3\mathbf{j})$
   (c) 52 J   (d) 52 J   (e) 26 W

Exercise 10c – p. 221
1. (a) -10 mph   (b) 90 mph
2. (a) -22 mph   (b) 42 mph
3. (a) $13\mathbf{i} + 19\mathbf{j}$ (b) $-13\mathbf{i} - 19\mathbf{j}$
4. (a) $18\mathbf{i} - \mathbf{j}$ (b) $5\sqrt{13}$ units
5. $9\mathbf{i} + 13\mathbf{j}$
6. (a) $8\mathbf{j}$; $\frac{5}{\sqrt{2}}(\mathbf{i} - \mathbf{j})$ (b) $\frac{5}{\sqrt{2}}\mathbf{i} + (8 - \frac{5}{\sqrt{2}})\mathbf{j}$
7. (a) $2\mathbf{i} + 6\mathbf{j}$ (b) $5\mathbf{i} + \mathbf{j}$
   (c) $\mathbf{i} - 4\mathbf{j}$

Exercise 10d – p. 222
1. (a) $3\mathbf{i} - 4\mathbf{j}$ (b) $-3\mathbf{i} + 4\mathbf{j}$
2. (a) $-10\mathbf{i} + 4\mathbf{j}$ (b) $10\mathbf{i} - 4\mathbf{j}$
3. (a) $-11\mathbf{i} + 10\mathbf{j}$ (b) $11\mathbf{i} - 10\mathbf{j}$
4. (a) $12\mathbf{i} + 8\mathbf{j}$ (b) $-12\mathbf{i} - 8\mathbf{j}$
5. (a) $P\!:\!t(6\mathbf{i} - \mathbf{j})$, $Q\!:\!t(3\mathbf{i} + 7\mathbf{j})$
   (b) $-9\mathbf{i} + 24\mathbf{j}$
6. (a) (i) $t(10\mathbf{i} + 4\mathbf{j})$
      (ii) $36\mathbf{i} + 2\mathbf{j} + t(-8\mathbf{i} + 3\mathbf{j})$
      (iii) $36\mathbf{i} + 2\mathbf{j} + t(-18\mathbf{i} - \mathbf{j})$
   (b) they collide
7. (a) $\mathbf{v} = \frac{1}{4}t^2\mathbf{i}$, $\mathbf{r} = \frac{1}{12}t^3\mathbf{i}$
   (b) $\mathbf{v} = 3\mathbf{i} + 4\mathbf{j}$, $\mathbf{r} = 3\mathbf{i} + t(3\mathbf{i} + 4\mathbf{j})$
   (c) $-\mathbf{i} + 4\mathbf{j}$ (d) $\frac{29}{3}\mathbf{i} + 16\mathbf{j}$
8. (a) $10\mathbf{i} + t(12\mathbf{i} + 5\mathbf{j})$
   (b) $20\mathbf{i} - 4\mathbf{j} + t(-3\mathbf{i} + 10\mathbf{j})$
   (c) $-10\mathbf{i} + 4\mathbf{j} + t(15\mathbf{i} - 5\mathbf{j})$
   (d) 2.40 pm
9. (a) $\mathbf{i} - 2\mathbf{j} + 4t\mathbf{i} + 3t^2\mathbf{j}$ (b) $\sqrt{181}$
   (c) $(2t^2 + t - 1)\mathbf{i} + (t^3 - 2t)\mathbf{j}$
   (d) 29 km

CHAPTER 11

Exercise 11a – p. 229
1. (a) 120 N s   (b) 24 000 N s
   (c) 11 040 N s   (d) 1177×10^4 N s
   (e) 4 N s
2. (a) 84 N s   (b) 72×10^4 N s
   (c) 88 N s   (d) 6000 N s ($g = 10$)
3. 2
4. -2
5. 25
6. -2
7. $\frac{8}{3}$
8. 3 s
9. (a) $26\mathbf{i}$ (b) $-6\mathbf{i}$
10. 7 N
11. 32 N
12. (a) $\frac{76}{5}\mathbf{i}$ (b) $12\mathbf{i}$; 17.5 s, 22.5 s
13. (a) 160 N   (b) 20 N
14. (a) 31 200 N s   (b) 31 200 N s
   (c) 15 600 N

Exercise 11b – p. 231
1. 12.1 N s
2. 10.2 N s
3. 12.5 N s
4. 0.72 N s
5. 108 N s

Exercise 11c – p. 235
1. 3 m s^{-1}
2. 0
3. -1.5 m s^{-1}
4. 0
5. 6 kg
6. 1.5 kg
7. 3 m s^{-1}
8. 6 m s^{-1}
9. $u = 2$, $v = 7$
10. $6\frac{2}{3}$ m s^{-1}
11. 1 kg
12. (a) $\sqrt{2gl}$ (b) $\frac{1}{2}\sqrt{2gl}$

Exercise 11d – p. 238
1. (a) (i) 12 J   (ii) 6 N s
   (b) (i) 58.5 J   (ii) 18 N s
   (c) (i) 240 J   (ii) 30 N s
   (d) (i) 96 J   (ii) 24 N s
   (e) (i) 45 J   (ii) 30 N s
   (f) (i) 3.6 J   (ii) 1.8 N s
2. $13\frac{1}{3}$ N s; 4 N s
3. (a) 600 kg   (b) 480 N s   (c) 480 J
4. (a) (i) u (ii) $3mu$ (iii) $6mu^2$
   (b) (i) $3u$ (ii) $3mu$ (iii) $6mu^2$

5. (a) $2\sqrt{10}$ m s^{-1} $(6.3$ m s$^{-1})$
   (b) $6\sqrt{10}$ N s $(19$ N s$)$
   (c) $2\sqrt{7}$ m s^{-1} $(5.3$ m s$^{-1})$
   (d) $6(\sqrt{10}+\sqrt{7})$ N s $(35$ N s$)$

CONSOLIDATION C

Miscellaneous Exercise C – p. 241

1. 37.6 N s
2. (a) $6\mathbf{j}$ (b) $\frac{3}{2}(9+4t^2)$
   (c) $7\mathbf{i}+5\mathbf{j}$
3. (a) 4 m s^{-1} (b) 3 J
4. (a) $\mathbf{v}=10\mathbf{i}+2t\mathbf{j}$ (b) 10.8 m s^{-1}
5. (a) $\dfrac{7}{g}$ s (0.714) and $\dfrac{35}{g}$ s (3.57)
   (b) $\to 10$ m, $\uparrow 12.5$ m; $\to 50$ m, $\uparrow 12.5$ m
6. 5.88 N s
7. 3.05 s, 18.5 m
8. A
9. D
10. D
11. C
12. (a) 20 m s^{-1} (b) 2.16 N
13. (a) 4.5 m s^{-1} (b) 6750 J
   (c) 2250 N s
14. (a) 10 m   (b) 2 s   (c) 24.9 m s^{-1}
15. (a) 48 m   (b) 116 m   (c) 49.5 m s^{-1}
16. (a) 56 m s^{-2} (b) 156 m
   (c) $6\mathbf{i}-8\mathbf{j}$ (d) $\mathbf{i}+8\mathbf{j}$
   (e) $\sqrt{65}$ m s^{-1} (8.06) (f) $82.9°$
17. (a) 3 s   (b) 16 m   (c) 0.75 rad
18. $H=\dfrac{V^2}{2g}$; $T=\dfrac{V}{g}$; $V=\frac{3}{8}U$
19. (a) 265 km h^{-1} (b) $344°$
   (c) $43:53$ $(1:1.23)$
20. (b) $78°$, $10°$ (c) 2 s
21. 125 s, 100 s, 300 m
22. (b) $\mathbf{i}-\mathbf{j}$ (c) 5 m s^{-1} (d) 25 J
23. (a) 16.6 m   (b) Yes, by 0.11 m
24. 29.5 s
25. (b) 36.75 m
26. $2m\sqrt{\dfrac{2gh}{3}}$
27. 0.0075 s
28. (b) 2.86 s   (c) 17.5 m s^{-1}
29. 0.036 N s
30. (a) $\dfrac{d\mathbf{v}}{dt}=(2t-1)\mathbf{i}+(2-2t)\mathbf{j}$
   (b) $\dfrac{d^2\mathbf{v}}{dt^2}=2\mathbf{i}-2\mathbf{j}$
   (c) $8t^2-12t+5$

31. A : $(40t^2+0.5)\mathbf{i}$;
   $_A\mathbf{r}_B=(40t^2-0.25)\mathbf{i}-30t^2\mathbf{j}$;
   $T=\frac{1}{50}\sqrt{10}$ s
   $F_1:4$ J, $F_2:2.4$ J

CHAPTER 12

Exercise 12a – p. 251

1.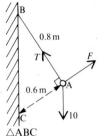
$\triangle ABC$
$F=6$ N, $T=8$ N

2.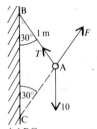
$\triangle ABC$
$F=10\frac{\sqrt{3}}{3}$ N $=T$ $(5.77$ N$)$

3.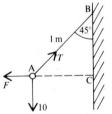
$\triangle ABC$
$F=10$ N, $T=10\sqrt{2}$ N $(14.1$ N$)$

4.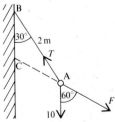
$\triangle ABC$
$F=10$ N, $T=10\sqrt{3}$ N $(17.3$ N$)$

5.

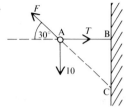

$\triangle ABC$
$F = 20$ N, $\ T = 10\sqrt{3}$ N $\ (17.3$ N$)$

6.

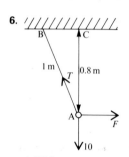

$\triangle ABC$
$F = 7\frac{1}{2}$ N, $\ T = 12\frac{1}{2}$ N

7.

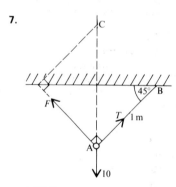

$\triangle ADC$
$F = 5\sqrt{2}$ N $= T$ $\ (7.07$ N$)$

8.

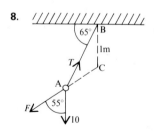

$\triangle ABC$
$T = 18.1$ N, $\ F = 10$ N

Exercise 12b – p. 253

	P	Q
1.	34.6	40
2.	11.3	7.99
3.	3.54	4.34
4.	8.51	2.91
5.	67.2	82.4
6.	17.1	11.1

Exercise 12c – p. 254
1. $P = 12.5$, $\ Q = 7.5$
2. $P = 8\sqrt{3}$ (13.9), $\ \theta = 30°$
3. $\theta = 154°$, $\ P = 18.4$

Exercise 12d – p. 259
1. (a) 12 N m $\circlearrowright$ (b) 19.2 N m $\circlearrowleft$
    (c) 10.5 N m $\circlearrowleft$ (d) 22 N m $\circlearrowleft$
2. (a) 11 N m $\circlearrowleft$ (b) 1.5 N m $\circlearrowleft$
    (c) 3 N m $\circlearrowleft$ (d) 0
    (e) 2.5 N m $\circlearrowleft$ (f) 0
    (g) 32 N m $\circlearrowleft$
3. (a) 1 N m $\circlearrowright$ (b) 2.53 N m $\circlearrowright$
4. (a) $\sqrt{3}$ N m $\ (1.73$ N m$)$
    (b) $(3\sqrt{3}+2)$ N m $\ (7.20$ N m$)$
5. (a) 5.5 N m   (b) 2.75
6. 10 N m
7. (a) 23 N m $\circlearrowleft$ (b) 6 N m $\circlearrowleft$

CHAPTER 13

Exercise 13a – p. 263
1. A: $11\frac{2}{3}$ N,   B: $13\frac{1}{3}$ N
2. A: 18 N,   B: 17 N
3. A: 26 N,   B: 6
4. A: 36 N,   B: 0
5. 0.43 m;   500 N
6. $T_1 = \frac{2}{3}W$, $T_2 = \frac{1}{3}W$
7. $T_1 = \frac{3}{2}W$, $T_2 = 0$
8. $T_1 = 2\frac{1}{3}W$, $T_2 = 4\frac{2}{3}W$
9. $T_1 = 0$, $T_2 = 3W$
10. 1.03 m
11. (a) 150 N, 570 N
    (b) 312 N, 408 N (3 sf)
    (c) each 360 N
12. 1.375 m
13. (a) 54.3 kg   (b) 10.9 kg
14. $3W$
15. (a) 0.825 m   (b) 0.8 m
16. (a) $\frac{3}{4}a$ (b) $\frac{3}{4}$
17. (a) 0.96 m

Exercise 13b – p. 269

1. $34°$
2. (a) 200 N (b) 100 N (c) $\frac{1}{2}$
3. $28°$
4. $120\sqrt{3}$ N $(208$ N$)$; $\mu = \frac{3}{7}\sqrt{3}$ (0.742)
5. To the top
6. $21°$
7. 12 N; 12 N↑
8. (a) 12 N (b) 20.8 N
9. $12\sqrt{3}$ N (20.8)
10. (a) 24 N (b) 24 N at $60°$ to PR
11. (a) $53\frac{1}{3}$ N (b) 44.3 N at $16°$ to PQ
12. (a) $W\frac{\sqrt{3}}{3}$ (b) $W\frac{\sqrt{3}}{6}$
13. (a) 1.125 (b) $26\frac{2}{3}$ N
14. (a) 13.5 N (b) $9\sqrt{3}$ N (15.6)

Exercise 13c – p. 274

1. (a)

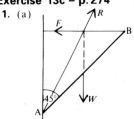
  (b) $R = \frac{1}{2}W\sqrt{5}$ $(1.12W)$
2. (a)
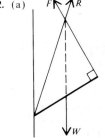
  (b) $R = \frac{1}{4}W\sqrt{7}$ $(0.661W)$
3. (a)
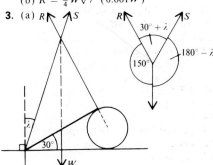

(b)
$$\frac{R}{\sin(180° - \lambda)} = \frac{S}{\sin 150°} = \frac{W}{\sin(30° + \lambda)}$$

5. $56°$: $\frac{3}{4}W$
6. $\sqrt{2} - 1$
7. $53°$
8. There must be friction at the ground to counteract the horizontal reaction at the wall.
9. (a)
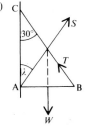

  (b) $\lambda = 30°$; $\mu = \frac{1}{3}\sqrt{3}$ (c) $\frac{1}{3}W\sqrt{3}$
10. (a) The reaction R and the weight meet at the centre O; the tension must be concurrent with R and W.
  (b) AO = 3 m, CO = 5 cm, CA = 4 cm
  (c) $\triangle$CAO; $T = 25$ N, $R = 15$ N

CHAPTER 14

Exercise 14a – p. 280

1. (a) 16 cm (b) 6 cm
  (c) 17.5 cm (d) 9.6 cm
  (e) 18 cm (f) 10.3 cm
2. $\frac{40}{9}a$ $(4.44a)$
3. $\frac{5}{2}a$ $(2.5a)$
4. $\frac{11}{5}a$ $(2.2a)$
5. $\frac{13}{5}a$ $(2.6a)$
6. $\frac{7}{4}a$ $(1.75a)$
7. $\frac{34}{11}a$ $(3.09a)$
8. $3.9a$
9. $3.5a$
10. $5a$
11. (a) 3 (b) $(\frac{10}{3}, 0)$
12. (a) 34.7 cm (b) 0.8 kg
  (c) 36.7 cm

Exercise 14b – p. 283

1. $\left(\frac{13}{5}, \frac{16}{5}\right)$
2. $\left(\frac{6}{7}, \frac{16}{7}\right)$
3. $\left(\frac{1}{5}, \frac{11}{5}\right)$
4. $\left(-\frac{4}{5}, \frac{6}{5}\right)$
5. $\left(5, \frac{23}{5}\right)$
6. $\left(\frac{1}{3}, 1\right)$

Exercise 14c – p. 289

1. (a) (2, 2)   (b) (4, 3)
   (c) (2, 1)   (d) (2, 1)
2. (a) (3.5, 5.5), (7, 4)
   (b) (1.5, 3.5), (5, 1.5)
   (c) (3, 3), (6, 4)
   (d) (3, 3), (6, 4)
   (e) (3, 4), (7, 4.5)
   (f) (2, 4), (4, 2), (6, 4)

3.

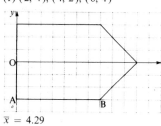

$\bar{x} = 4.29$
$\bar{y} = 0$

4.

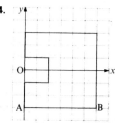

$\bar{x} = 3.25$
$\bar{y} = 0$

5.

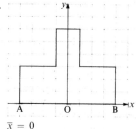

$\bar{x} = 0$
$\bar{y} = 2.1$

6.

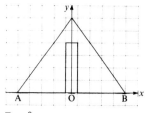

$\bar{x} = 0$
$\bar{y} = 2$

7. $\left(\frac{13}{3}, \frac{7}{3} \right)$

8. $\left(\frac{113}{24}, \frac{79}{24} \right)$

9. (5.6, 4.3)

10. (5.5, 3)

11. $\left(\frac{16}{3}, \frac{14}{3} \right)$

12. $\left(\frac{119}{26}, \frac{58}{13} \right)$

Exercise 14d – p. 295

1. 35°
2. 43°
3. 28°
4. 39°
5. 55°
6. $\frac{48}{23}$ and $\frac{96}{23}$
7. (a) $\bar{x} = \dfrac{4a^2 + 3ad}{4a + 2d}$, $\bar{y} = \dfrac{2a^2 + 2ad + d^2}{2a + d}$
   (b) $d = \frac{1}{3}a$

Exercise 14e

1. 45°
2. (a) 59°; $\frac{5}{3}$ (b) 310; $\frac{3}{5}$
3. 0.8
4. (a) 27°   (b) 0.1

CHAPTER 15

Many of the problems do not have 'correct answers'. There are many possible responses and in most cases the following factors are relevant; the value used for g, objects treated as particles, accuracy of measurements. Some of the other, less obvious, responses are given here.

Also, to save space, the answers are given in note form – it is better practice to write out reasons as proper sentences.

Exercise 15a – p. 299

1. Wind speed, air resistance ignored. Brick, no initial velocity assumed.
2. Cyclist and cycle treated as particle. Assumed: no frictional resistance in cycle, no loss of effort in machine, constant resistance to motion.
3. Work done by hoist in raising itself ignored.
4. Air resistance ignored. The value of 2.3 is not reliably accurate to 2 sf because of simplification, including rounding errors accumulated from, e.g., using only 2 sf for the value of g.
5. Assumed: no friction between rope and pulley, no air resistance, rope does not stretch. Radius of winch, weight of rope ignored.

6. String horizontal. No friction between string and table edge.

7. Air resistance reduces distance. Wind increases or decreases distance depending on direction.

8. Fix vinyl floor covering to horizontal surface. Attach spring balance to disc and pull until disc just moves. Use reading on balance (convert to newtons if necessary) as value of limiting frictional force. Calculate μ using $F = \mu R$. Assume μ constant over whole surface, limiting dynamic friction equal to limiting static friction.

9. Measure time taken by car to cover measured short distance from rest at maximum power. Calculate acceleration, final velocity, hence driving force, $\Rightarrow$ power ($=$ driving force $\times$ velocity). Assumptions: negligible resistance, acceleration constant.
   Check: On slope with variable inclination, find maximum gradient car can just climb.
   Use driving force $=$ component of weight down plane.
   Further assumption: no slipping occurs.

10. Resolve and take moments to give three equations from which θ can be found. Assume man is a point load at top of ladder. Result tends to err on side of safety as person cannot stand vertically at top of ladder (i.e. at point of contact of ladder and wall): taking moments about top of ladder gives equation of form
$$\tan \theta = \frac{A}{B - kW}$$ where A, B and W (person's weight) are known as k is distance from top of ladder of point through which person's weight acts; as k increases, $\tan \theta$ and hence θ increase.

11. Use firing speed to calculate impulse from change in momentum. Assume: 'bird' does not bounce off screen but stops dead, screen does not break, 'bird' hits screen at speed with which it is fired.

12. Using data, average acceleration over each distance interval can be calculated. A 'picture' of motion can be obtained by plotting data on distance-time graph. Motion can be modelled by trying to 'fit' a curve to points and using curve as equation of motion.

13. Measure distance bullet penetrates into wood, then find deceleration from $v^2 = u^2 + 2as$. Assume: block fixed, muzzle speed of gun known, deceleration constant.

14. Wind resistance greater, engine not tuned for maximum speed, poor surface, inexperienced driver.

15. (a) Resistance to motion at maximum speed down slope is equal to that for maximum speed on level.
    (b) Component of car's weight down steep hill is large compared to driving force at high speeds (as $P = Dv$, for constant P, D decreases as v increases); resistance increases with speed, but speed would have to be very high for resistance to equal combined effect of driving force and weight, so high that maximum engine revs would probably be exceeded and engine destroyed.

16. About 220 m. Air resistance is significant, wind effect is unpredictable during 'drop', instruments measuring plane's speed and height are probably not reliably accurate enough.

Exercise 14e
1. $45°$
2. (a) $59°$; $\frac{5}{3}$ (b) 310; $\frac{3}{5}$
3. 0.8
4. (a) $27°$ (b) 0.1

CONSOLIDATION D

Miscellaneous Exercise D – p. 305
1. B
2. A
3. B
4. (a) 784 N   (b) 0.5 m
5. 44 Nm in the sense BAD, 27 N, -9 N
6. P: $13\frac{1}{3}$ N,   Q: $26\frac{2}{3}$, 20 N
7. (a) $8\mathbf{i} + 6\mathbf{j}$, $2\mathbf{j}$ (b) $8\sqrt{2}$ N
    (c) 16 Nm clockwise
8. 3.5 m
9. AD: 5 N,   DC: 20 N,   CA: 35.4 N
10. $3\mathbf{i} + 2.5\mathbf{j}$

11. 0.18
12. (a) 139 N   (b) 18.9 N
13. $R - P = 2S \tan\theta$, $R + P = W$
    (a) $\tan\theta = \frac{1}{2}$
14. $2.42a$, $2.42a$
15. $\frac{1}{3}\sqrt{3}$
16. (a) $\frac{3}{5}d$ (b) d (c) $59°$
17. F
18. T
19. F
20. F
21. (b) $\frac{8}{9}a$
22. (a) 2 cm   (b) $1\frac{2}{3}$ cm   (c) $40°$
24. 0.346
26. $\dfrac{W(a\sin\theta - b)}{b\cot\theta}$, $\dfrac{W a \sin^2\theta}{b}$;

    end A is about to slip up the wall

INDEX